The Truth Behind "free energy"

Published by Spoo Publications Inc.

Enter the world of "free energy" find the truth behind the lies, the facts behind the fluff and find out how an ordinary person can create the power needed to support a family or fuel a revolution!

The End of power slavery is in our hands right now, how long with big industry keep us chained to the wall of servitude? How long will YOU keep paying for what is and can be free?

Would you pay double or triple value for a free car? Hell no! exactly, but that's what you're doing every day when you buy power off the grid.

Norman Wootan is an open source engineer who has been helping the public access the suppressed and alternative energy field for a number of years now. Norm's efforts include experience and insights into resurrecting the publicly suppressed EVGRAY energy technology. Also with his associate Joel the magnetic resonance amplifier. Both these accomplishments are of considerable value to aid the task of bringing education about free energy technology into public hands.

Norm's work in progress on the

E.V. Gray skewed repulsion motor

Norman Wootan along with Mark McKay,Peter Lindemann, John Bedini, Alan Francoeur and Jerry Decker have all done extensive research into the inner workings of the suppressed EVGRAY energy motor systems.

The now murdered EVGray left center, with his energy

FREE ENERGY system being tested an analyzed.

The EVgray Motor systems are capable of FREE ENERGY, the process functions by the ability to run an engine from a battery source, then convert/transform this energy into higher power.

This energy transformation process operates the motor and keeps the batteries recharged. This process is interpreted as a utilization and transformation of radiant energy or 'cold electricity'. This is also rationalised as the same principles first discovered by Nikola Tesla and used in his Magnifying Transmitter.

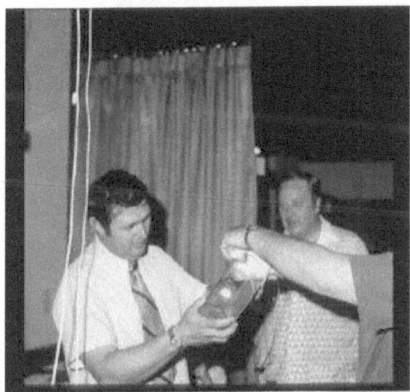

Taken from <u>Here</u>

Above is a light bulb being lit under water from the Radiant Energy/Cold Electricity. Please visit Peter Lindermen's site for the complete details. Further background on cold current as described by Dr Lindermen can be found in Secrets of Cold War Technology: Project HAARP and Beyond (Paperback) by Gerry Vassilatos.

Note- This book is out of print. Panacea has a back up copy available only for educational purposes and reference, to acquire a copy for evaluation, you must show Panacea that you require it as a non profit goal and will not breach the copyright laws. Upon our decision you may attain a copy via electronic (PDF) format.

For further references, visit the peswiki.com website. Here is an extract taken from the page describing both videos Norm and Dr lindermen made.

N. Wootan - History of The E.V. Gray Motor (Free Energy - KeelyNet Conference) The truth is finally revealed. The mystery about Ed Gray's EMA motors is finally over. With two recovered EMA motor prototypes on stage at the KeelyNet Conference in June 2001, Norman Wootan discusses every design feature possible. Every single way the real motors deviate from the designs revealed in Gray's Patent are discussed in detail.

Now you can see with your own eyes how it was really done. This video is a must for serious researchers wanting to convert Radiant Energy into mechanical power. A great companion piece to "The Free Energy Secrets of Cold Electricity" by Dr. Lindermen (book or video) where the EMA power supply is discussed.

Peter Lindemann speaking about the EVGRAY suppression

and technical aspects related to the motors.

Peter Lindermen - The Free Energy Secrets of Cold Electricity (KeelyNet Conference). In the 1970's, inventor Edwin Gray developed an electric automobile engine that produced 80 horsepower and recharged its own batteries. It ran on what he called "cold electricity." This amazing technology remained shrouded in mystery until September 2000.

This 3 part video is the complete technical lecture given by Dr. Lindemann at that time. In it, he explains exactly how Ed Gray's system works, how he produced "cold electricity" and how that relates to Nikola Tesla's earlier discovery of "Radiant Energy". Using 50 slides of articles, patents, photos, and circuit diagrams, Dr. Lindemann documents his research, until the method is fully revealed. Now you can understand one of the most powerful Free Energy methods ever discovered. There is enough information presented here so that any competent electrical technician can duplicate this process.

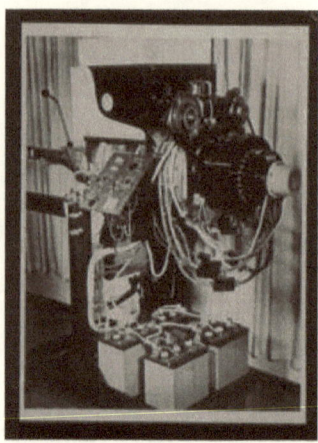

Taken from Peter Lindemann's site. Above is the Gray motor

prototype which never made it into public hands.

This motor is described by Dr Lindermen as: Quote- In 1958, Edwin V. Gray, Sr. discovered that the discharge of a high voltage capacitor could be shocked into releasing a huge, radiant, electrostatic burst. This energy spike was produced by his circuitry and captured in a special device Mr. Gray called his "conversion element switching tube." The non-shocking, cold form of energy that came out of this "conversion tube" powered all of his demonstrations, appliances, and motors, as well as recharged his batteries. Mr. Gray referred to this process as "splitting the positive."

During the 1970's, based on this discovery, Mr. Gray developed an 80 hp electric automobile engine that kept its batteries charged continuously. Hundreds of people witnessed dozens of demonstrations that Mr. Gray gave in his laboratory.-End quote

So far only independent engineers have committed time and FINACIAL resources towards uncovering the EVGray energy system for public benefit. The scientific community have no prior knowledge or faculty information which understands or teaches about the COLD CURRENT energy principles used in the suppressed EVGray energy systems.

Mark McKay, Alan Francoeur, Peter Lindermen, Jerry and Norm have so far been the only individuals to publicly disclose the information open source and commit scientific investigation into the suppressed EVGray energy systems.

Norm speaking at a KeelyNet energy conference describing the

suppressed EVGRAY motors

With out Alan, Norm and Dr Lindermen there would be no public knowledge of the suppressed EVgray systems, can you reason now why for security reasons an independent research and development centre is essential to protect and understand this technology?

EVGray's motors can provide FREE ENERGY to run your car or house, it is no wonder he was killed and there is virtually no trace or public knowledge of EVgray's systems except for the information retrieved by the individuals mentioned.

Various open source and proprietary engineers have since followed these findings and offered independent replications for further study. At this time of 2008 none of them are at the same performance levels as the original EVGRAY motors

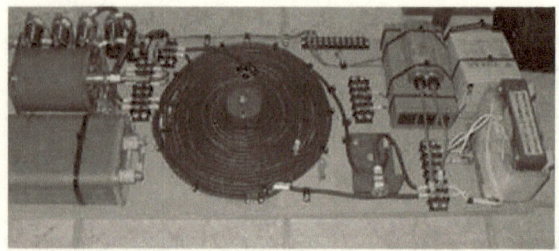

Open source R & D based on the findings of the EVGRAY systems.

Taken from Pure energy systems.

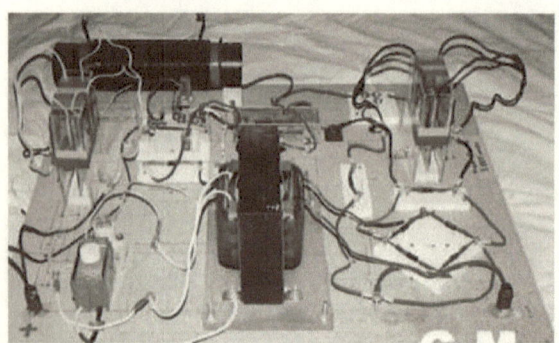

An attempted reproduction of the suppressed EVgray energy system submitted

taken from Pure energy systems.

Please visit these pages for Technical detail of the Gray systems:

With consolidation of these experienced engineers along with Panacea's public granted resource support, these principles can be perfected and delivered into further public benefit.The suppressed EVGray systems are one of the most unique and promising neglected alternative energy devices.

This research and development critically needs public granted supported resources in order to enable further scientific investigation by the experienced engineers. Please help and donate to Panacea!

Norm and Joel's SUPPRESSED MRA (magnetic resonant amplifier)

Quote-It was felt that this information should be released to the public as rapidly as possible, rather than risk loss of the information or the demo by 'circumstances' beyond our control.

There have been too many discoveries which have disappeared by not being openly shared. We would rather risk being 'flamed' by other experimenters who actually BUILD THE CIRCUIT than take the chance of the information being lost by keeping it within a select group. Our two primary fears are that someone will attempt to patent what is intended to be a gift to humanity, possibly with an intent to profit from others work or to lock it away, the other that it might be suppressed in some other fashion.

Therefore, the widest possible distribution is requested, and duplication attempted wherever possible. At the time of this correction to the original file, we have confirmation as to the numbers and others are working on the circuit on their own. Please take this information in the spirit of how it is given, as a gift to humanity. The inventors are Joel McClain and Norman Wootan. You can contact them via KeelyNet or directly -Joel McClain & Norman Wootan -End quote

Norm has a signed affidavit from Dr. Bass which was submitted to the Patent Office for their patent application. This statement confirmed by DR Bass showed a 256:1 energy gain. This is quite a significant Over Unity, to put it mildly.

Quote - The Patent Office rejected the patent with no explanation (which is illegal) .When challenged by Dr. Bass the Patent Office simply said "sue us". Norman Wootan -End quote

Norm Giving evidence of the Patent experience on the ABC

documentary A Machine to Die For

Another quote relating to the suppression of the MRA from Tom Beardon. Quote- McLain and Wooten patented a great little MRA (Magnetic Resonance Amplifier) system, based on that application. Dr. Robert Bass, a very fine electrodynamicist of exceptional knowledge, experience, and ability wrote the patent for them, and assisted in their work.

For that he was persecuted, unjustly attacked, and suffered financial difficulties. The "system" does not forgive highly qualified scientists who take a serious interest in "perpetual motion machines" -- as permissible Maxwellian open dissipative systems are erroneously and derogatorily labeled by the orthodox scientific community. Any scientist violating that inquisition suffers the consequences. -End quote

Further suppression details of the MRA are profiled further in the Panacea energy suppression section.Norm's and Joel's work cannot progress in the current cartel revenue dominated environment.

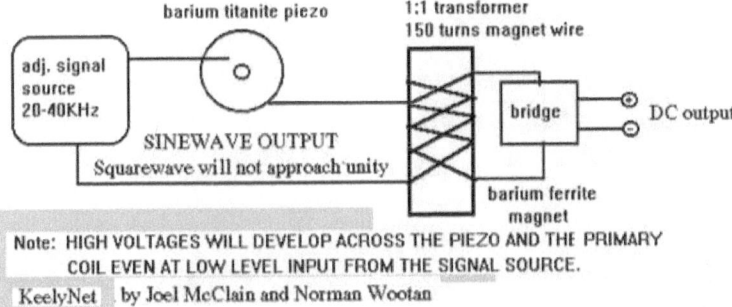

adj. signal source 20-40KHz

barium titanite piezo

1:1 transformer
150 turns magnet wire

bridge

DC output

SINEWAVE OUTPUT
Squarewave will not approach unity

barium ferrite magnet

Note: HIGH VOLTAGES WILL DEVELOP ACROSS THE PIEZO AND THE PRIMARY
COIL EVEN AT LOW LEVEL INPUT FROM THE SIGNAL SOURCE.
KeelyNet   by Joel McClain and Norman Wootan

A technical Schematic of the MRA principle courtesy of Jerry Decker's KeelyNet.

Another Quote from Norm taken from Keely Net: In regards to the MRA I would like to have answers to the many lurking questions. Every now and then I fire on up and play with it. It is real interesting to watch the current into an amplifier driving the MRA at max gain actually fall far below idle current draw. Hmmm! Want to have a real laugh? I bought a 100 watt solid state power amp from Radio Shack and purchased the life time warranty. The first time I put it to a real power test in the MRA circuit this amp literally smoked.

When I received it back from the Tandy repair facility it had a note attached. "We cannot understand how this amp was totally destroyed when it is very well fuse protected". Three times Tandy repaired this amp. Each time they had to replace all the transistors and the output transformer.

We posted many warnings to researchers about the hazards to solid state equipment when trying to measure a circuit running at resonance at extremely high "Q". Oh! by the way the MRA operates exactly like a well tuned Tesla coil. The voltage and energy multiplication is accomplished by Q ratio and not turns ratio. They do not comply with conventional transformer theory. These are extremely non-linear devices therefore display unusual performance. -End quote

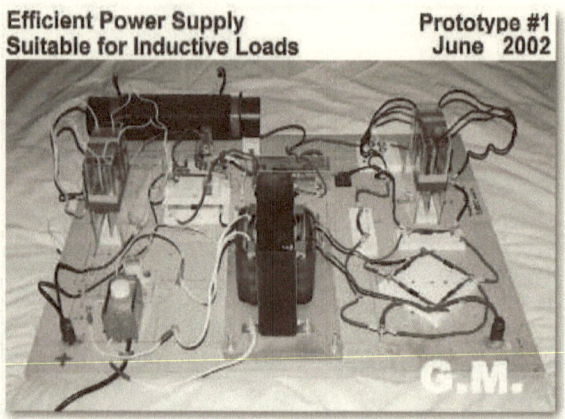

Efficient Power Supply
Suitable for Inductive Loads

Prototype #1
June 2002

G.M.

click image for full-size

Image Description by Sterling D. Allan
May 29, 2004

Working device by G.M. Close to Gray's patent 4,595,975, except that Gray used a "closed tube," while G.M. uses an "open air" tube. G.M. will also be specifying exact components, whereas Gray's patent is not that specific. G.M. had to make a "wild guess" on this prototype, based on a Technical Report by Richard Hackenberger and an understanding of the "high power pulse" principle.

G.M. said that this power supply output was "too powerful for bench top type work," so he scaled it down in prototype 2.

The above first prototype was disassembled for space considerations to make way for the next iteration. The second, smaller prototype is presently working. Images pending.

Prototype 1 used purely opposing permanent magnets. Prototype 2 uses a combination of electromagnets and permanent magnets.

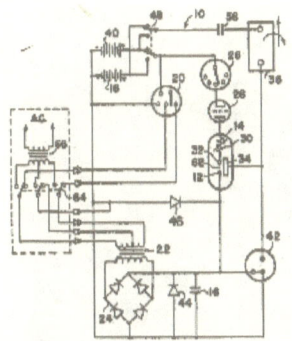

Pulsed Electromagnetic Motor by G.M. -- Prototype 2

Presently-working device by G.M. is a variation of Ed Gray's patent 3,890,458. G.M.'s version (variant) uses a combination of permanent magnets and electromagnets.

click image for full-size image

click image for full-size image

click image for full-size image

click image for full-size image

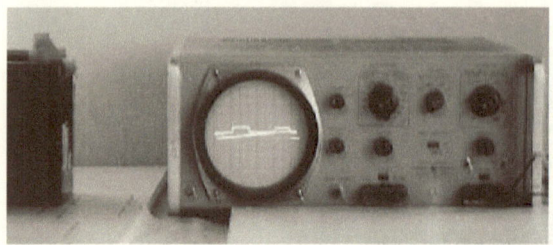

click image for full-size image

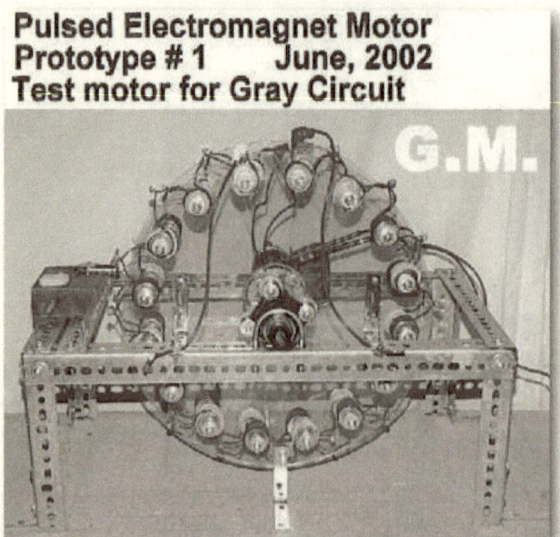

Pulsed Electromagnet Motor
Prototype # 1   June, 2002
Test motor for Gray Circuit

G.M.

click image for full-size image

Working device by G.M. is a variation of Gray's patent 3,890,458. Uses permanent magnets.

The above first prototype was disassembled for space considerations to make way for the next iteration. The second, smaller prototype is presently working. Images pending. May 29, 2004.

Nikola Tesla was a multi-disciplinary genius. His discovery of the rotating magnetic field in 1882 lead to a series of US Patents in 1888, which gave us the AC electric power system still in use today. This one achievement earned him the honor of being called "The Man Who Invented the 20th Century".

But his research went way beyond what has found its way into everyday use. He is the recognized inventor of the brushless AC induction motor, radio, remote control by radio, super-conductivity, fluorescent lighting, the bladeless turbine engine and pump, the capacitor discharge ignition system for automobile engines, the mechanical oscillator, and dozens of other inventions. But he also discovered that useful energy could be extracted from the heat of the ambient air, and that electric power in the form of Radiant Energy could be broadcast to everyone in the world through the ground.

In his masterful article _The Problem of Increasing Human Energy_, first published in Century Illustrated Magazine in June 1900, Tesla discusses the "energy situation" like never before. After discussing every known method of gathering energy from the Natural World, Tesla departs into the unknown. His first discussion is about a machine that can gather heat from the ambient air. He calls it a "Self-acting Engine" since it could run indefinitely from the solar energy stored in the air. He called it "the ideal way of obtaining motive power".

Tesla worked for years trying to solve all of the technical issues presented by the idea. His work with liquified air, his discovery of super-conductivity at ultra-low temperatures, his bladeless turbine and mechanical oscillator were all _spin-offs_ from his work on the ambient air engine. He was convinced the system could work and that it was absolutely the best way to harness solar energy.

On a world that is warming up, tapping ambient sources of heat in the air, water and ground are the most important technologies to develop at this time. For a brief article on Tesla's amazing "Self-acting Engine".

But Nikola Tesla's most famous attempt to provide everyone in the world with free energy was his World Power System, a method of broadcasting electrical energy without wires, through the ground. His Wardenclyffe Tower, pictured above, was never finished, but his dream of providing energy to all points on the globe is still alive today

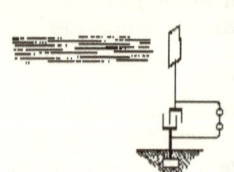

For starters, think of this as a solar-electric panel. Tesla's invention is very different, but the closest thing to it in conventional technology is in photo-voltaics. One radical difference is that conventional solar-electric panels consist of a substrate coated with crystalline silicon; the latest use amorphous silicon. Conventional solar panels are expensive, and, whatever the coating, they are manufactured by esoteric processes.

But Tesla's "solar panel" is just a shiny metal plate with a transparent coating of some insulating material which today could be a spray plastic. Stick one of these antenna-like panels up in the air, the higher the better, and wire it to one side of a capacitor, the other going to a good earth ground. Now the energy from the sun is charging that capacitor.

Connect across the capacitor some sort of switching device so that it can be discharged at rhythmic intervals, and you have an electric output. Tesla's patent is telling us that it is that simple to get electric energy. The bigger the area of the insulated plate, the more energy you get. But this is more than a 'solar panel' because it does not necessarily need sunshine to operate. It also produces power at night.

Of course, this is impossible according to official science. For this reason, you could not get a patent on such an invention today. Many an inventor has learned this the hard way. Tesla had his problems with the patent examiners, but today's free-energy inventor has it much tougher. At the time of this writing, the U. S. Patent Office is headed by a Reagan appointee who came to the office straight from a top executive position with Phillips Petroleum.

Tesla's free-energy receiver was patented in 1901 as An Apparatus for the Utilization of Radiant Energy. The patent refers to "the Sun, as well as other sources of radiant energy, like cosmic rays." That the device works at night is explained in terms of the night-time availability of cosmic rays. Tesla also refers to the ground as "a vast reservoir of negative electricity."

circuit controller

load

transformer

Tesla was fascinated by radiant energy and its free-energy possibilities. He called the Crooke's radiometer (a device which has vanes that spin in a vacuum when exposed to radiant energy) "a beautiful invention." He believed that it would become possible to harness energy directly by "connecting to the very wheelwork of nature." His free-energy receiver is as close as he ever came to such a device in his patented work.

But, on his 76th birthday, at the ritual press conference, Tesla (who was without the financial wherewithal to patent but went on inventing in his head) announced a "cosmic-ray motor." When asked if it was more powerful than the Crooke's radiometer, he answered, "thousands of times more powerful."

How it works

From the electric Potential that exists between the elevated plate (plus) and the ground (minus), energy builds in the capacitor, and, after "a suitable time interval," the accumulated energy will "manifest itself in a powerful discharge" which can do work. The capacitor, says Tesla, should be "of considerable electrostatic capacity," and its dielectric made of "the best quality mica,' for it has to withstand potentials that could rupture a weaker dielectric.

Tesla gives various options for the switching device. One is a rotary switch that resembles a Tesla circuit controller. Another is an electrostatic device consisting of two very light, membranous conductors suspended in a vacuum. These sense the energy build-up in the capacitor, one going positive, the other negative, and, at a certain charge level, are attracted, touch, and thus fire the capacitor. Tesla also mentions another switching device consisting of a minute air gap or weak dielectric film which breaks down suddenly when a certain potential is reached.

The above is about all the technical detail you get in the patent. Although I've seen a few cursory references to Tesla's invention in my sampling of the literature of free-energy, I am not aware of any attempts to verify it experimentally.

Plauson's converter

How it works

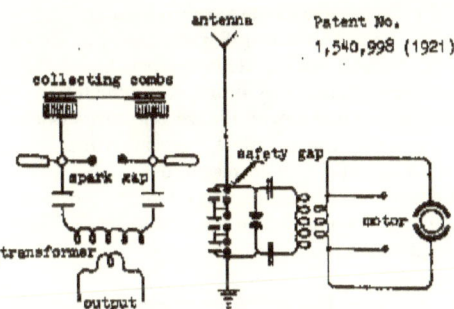

Plauson's converter

........If a single one-megaton nuclear warhead were exploded 300 miles over the center of the country, a high-voltage electromagnetic pulse would in theory disrupt communication and electrical systems all over the continental U.S. Gamma rays emitted by such an explosion would instantly strip away the electrons from air molecules in the upper atmosphere in roughly a circular, pancake-shaped zone. The free electrons would then accelerate radially with the earth's magnetic field, separating from the heavier, positively charged ions and creating a downward directed high-voltage electromagnetic pulse. This in turn results in electrical surges in all exposed conductors on the ground.

........When Nikola Tesla invented the AC (alternating current) induction motor, he had great difficulty convincing men of his time to believe in it. Thomas Edison was in favor of direct current (DC) electricity and opposed AC electricity strenuously. Tesla eventually sold his rights to his alternating current patents to George Westinghouse for $1,000,000. After paying off his investors, Tesla spent his remaining funds on his other inventions and culminated his efforts in a major breakthrough in 1899 at Colorado Springs by transmitting 100 million volts of high-frequency electric power wirelessly over a distance of 26 miles at which he lit up a bank of 200 light bulbs and ran one electric motor! With this souped up version of his Tesla coil, Tesla claimed that only 5% of the transmitted energy was lost in the process. But broke of funds again, he looked for investors to back his project of broadcasting electric power in almost unlimited amounts to any point on the globe. The method he would use to produce this wireless power was to employ the earth's own resonance with its specific vibrational frequency to conduct AC electricity via a large electric oscillator. When J.P. Morgan agreed to underwrite Tesla's project, a strange structure was begun and almost completed near Wardenclyffe in Long Island, N.Y. Looking like a huge lattice-like, wooden oil derrick with a mushroom cap, it had a total height of 200 feet. Then suddenly, Morgan withdrew his support to the project in 1906, and eventually the structure was dynamited and brought down in 1917.

........A Tesla coil is a special transformer that can take the 110 volt electricity from your house and convert it rapidly to a great deal of high-voltage, high-frequency, low-amperage power. The high-frequency output of even a small Tesla coil can light up fluorescent tubes held several feet away without any wire connections. Even a large number of spent or discarded fluorescent tubes (their burned out cathodes are irrelevant) will light up if hung near a long wire running from a Tesla coil while using less than 100 watts drawn by the coil itself when plugged into an electrical outlet! Since the Tesla coil steps up the voltage to such a high degree, the alternating oscillations achieve sufficient excitations within the tubes of gases to produce lighting at a minimal expense of original power! Fluorescent tubes can be held under high-tension wires to produce the same lighting up effect. Remember the farmer a few years ago who was caught with an adaptive transformer under a set of high tension lines that ran over his property? Through the air, he pulled down all the power he needed to run his farm without using any connecting apparatus to the lines overhead! Any electrical engineer with the proper materials can do the same thing.

........Incandescent bulbs burn high resistance filaments that gobble up energy. Fluorescent tubes burn filaments (cathodes) to create an electrical flow that sets their internal phosphorus coatings aglow. Using a Tesla coil, high voltage AC can light up glass-enclosed vacuum bulbs coolly without any gases inside them at all! Any number of cold light bulbs can be lit using only one Tesla coil, and since there is nothing inside them to burn out, they can last indefinitely. It seems like a low cost form of street lighting, doesn't it?

........When Tesla was determining the resonant frequencies of the earth to potentially transmit unlimited electric power, he also recognized frequencies that acted as a damping field to nullify electric power. With the advent of the wireless and Tesla's unique investigations into broadcasting electricity, a dozen or more inventors thereafter announced their own means for transmitting electrical energy without wires. One British inventor, H. Grindell-Matthews, actually demonstrated his "mystery ray" apparatus in 1924 to a Popular Science Monthly writer in London (See: Pop. Sci. Monthly, Aug. 1924, P. 33). When his beam was directed toward the magneto system of a gasoline engine, it stopped the system. Afterwards, it ignited gun powder, lit an electric lamp bulb from a distance and killed a mouse in seconds! Grindell-Matthews said the secret was involved with the "carrier beam" he used to conduct a high-voltage, low-frequency electrical current. During 1936, Guglielmo Marconi experimented with extremely low frequency (ELF) waves and displayed their exceptional ability to penetrate metallic shielding. These waves could affect electrical devices, overload circuits and cause machines like generators, electric motors and automobiles to stall. Diesel engines, which do not rely on electrical ignition, were not affected. Mysteriously, Marconi's research on the subject was never found after the war. A Cordless Future for Electricity?
John D. Sutter - CNN

09/02/2009

(CNN) -- Electronics such as phones and laptops may start shedding their power cords within a year. The original article on CNN.com is located here and will open in a new browser window.

Wireless electricity may soon make tangled power cords a thing of the past. That's the prediction of Eric Giler, CEO of WiTricity, a company that's able to power light bulbs using wireless electricity that travels several feet from a power socket.

GETTY IMAGES/FILE WiTricity's version of wireless electricity -- which converts power into a magnetic field and sends it sailing through the air at a particular frequency -- still needs to be refined a bit, he said, but should be commercially available soon.

Giler, whose company is a spinoff of a Massachusetts Institute of Technology research group, says wireless electricity has the potential to cut the need for power cords and throw-away batteries.

"Five years from now, this will seem completely normal," he said.

"The biggest effect of wireless power is attacking that huge energy wasting that goes on where people buy disposable batteries," he said. It also will make electric cars more attractive to consumers, he said, because they will be able to power up their vehicles simply by driving into a garage that's fitted with a wireless power mat.

Electric cars are "absolutely gorgeous," he added, "but does anyone really want to plug them in?"

Ideas about wireless electricity have been floating around the world of technology for more than a century. *Nikola Tesla* started toying with the ability to send electricity through the air in the 1890s. Since then, though, making wireless electricity safe and cheap enough to put on the market has been an arduous task for researchers.

Engineers have developed several ways to convert electricity into something that's safe to send through the air without a wire. Some of their technologies are available on commercial scales, but they have some limits.

Low-level power

One set of researchers is able to send power over long distances but in very small amounts.

For example, in 2003, a Pittsburgh, Pennsylvania, company called Powercast used radio waves to light a low-power LED bulb that was 1.5 miles from its power source, said Harry Ostaffe, spokesman for the company.

Now, Powercast's technology is used in office buildings to power temperature sensors that regulate air conditioning systems and in other low-power applications. The company also has sold wireless artificial Christmas trees strung with LED lights for about $400, Ostaffe said.

But radio waves can't transfer the larger amounts of electricity needed to power laptops or mobile phones, he said.

Power pads

Another type of wireless electricity technology can send large amounts of power over very small distances, often not more than a few centimeters.

Such technology is available today, but only in minimal ways. Think, for instance, about electric toothbrushes that sit on charging cradles but don't actually plug in.

One problem with the high-power, small-distance idea is that each device requires its own charging pad, and consumers hate that, said Menno Treffers, chairman of the steering group at the Wireless Power Consortium. The group formed in late 2008 to promote standardization of the technology.

Treffers said consumers soon should be able to buy one power pad that would charge all of their electronic devices. It might look like a placemat, and cell phones, remote controls and appliances would charge automatically when they're placed on the pad.

"The key reason to do it is convenience, because if you want to get rid of all the different power supplies, there are other ways that are cheaper," he said.

The pads, which would rely on electrical sockets as their initial sources of power, also would be more energy efficient than plugging all of the devices into power sockets directly, he said. The pads would shut off automatically when a device has finished charging and are about 70 to 90 percent as efficient as transferring power through a wire, he said.

Wire-free chargers for a single item are relatively cheap: about $10 to build, he said. But it's unclear how much pads that could power a living room worth of equipment would cost, he said.

'Magnetically coupled resonance'

Ultimately, Giler's group from MIT wants to combine the best of both worlds: large amounts of power sent over long distances.

Their technology is called "magnetically coupled resonance," and it basically sends a magnetic field through the air at a specific frequency that an an enabled phone or TV can pick up and turn back into electricity. It works kind of like sound. Think about how an opera singer can break a wine glass if he sings at just the right frequency.

Adding the technology to cell phones, mp3 players and other devices should not increase their cost much, he said.

Despite Giler's optimism, there are some doubts about magnetically coupled resonance.

Treffers said there may be health risks associated with the magnetic fields created in the MIT process. Giler said the technology would produce magnetic fields that are "about the same density as the Earth's magnetic field."

He said wireless electricity has many environmental benefits. Companies make about 40 billion disposable batteries each year, he said, and wireless electricity could do away with that.

The biggest barrier to the technology's adoption, he said, is that people just aren't familiar with the idea.

While in college Nikola Tesla claimed it should be possible to operate an electrical motor without sparking brushes. He was told by the professor that such a motor would require perpetual motion and was therefore impossible. In the 1880's he patented the alternating current generator, motor, and transformer.

During the 1890's he intensively investigated other methods of power generation including a charged particle collector patented in 1901. When the *New York Times* in June of 1902 carried a story about an inventor who claimed an electrical generator not requiring a prime mover in the form of an external fuel supply, Tesla wrote a friend that he had already invented such a device.

Fuelless electrical generation raises the same objection of perpetual motion as did the generator in use today when it was first proposed. Research Nikola Tesla carried out during his second creative period and the resulting devices that were the basis for his assertion of fuelless electrical generation will be examined. Whether Tesla's fuelless generator was a "perpetual motion scheme" of the sort his teacher warned him against, or a creative application of recognized natural phenomena will be discussed.

TESLA'S STATEMENTS

In *The Brooklyn Eagle*, Tesla announced, on July 10th, 1931, that "I have harnessed the cosmic rays and caused them to operate a motive device." Later on in the same article he said that "More than 25 years ago I began my efforts to harness the cosmic rays and I can now state that I have succeeded." In 1933, he made the same assertion in an article for the *New York American*, November 1st, under the lead in "Device to Harness Cosmic Energy Claimed by Tesla." Here he said:

This new power for the driving of the world's machinery will be derived from the energy which operates the universe, the cosmic energy, whose central source for the earth is the sun and which is everywhere present in unlimited quantities.

Dating back "more than 25 years ago" from 1933 would mean that the device Tesla was speaking about must have been built before 1908. More precise information is available through his correspondence in the Columbia University Library's collection. Writing on June 10th, 1902 to his friend Robert U. Johnson, editor of *Century* Magazine, Tesla included a clipping from the previous day's *New York Herald* about a Clemente Figueras, a "woods and forest engineer" in Las Palmas, capital of the Canary Islands, who had invented a device for generating electricity without

burning fuel. What became of Figueras and his fuelless generator is not known, but this announcement in the paper prompted Tesla, in his letter to Johnson, to claim he had already developed such a device and had revealed the underlying physical laws.

IDENTIFYING THE INVENTION

The device that, at first, seems to best fit this description is found in Tesla's patent for an "Apparatus for the Utilization of Radiant Energy," number 685,957, that was filed for on March 21, 1901 and granted on November 5, 1901. The concept behind the older technical language is a simple one. An insulated metal plate is put as high as possible into the air. Another metal plate is put into the ground. A wire is run from the metal plate to one side of a capacitor and a second wire goes from the ground plate to the other side of the capacitor. Then:

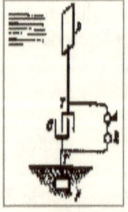
The sun, as well as other sources of radiant energy, throw off minute particles of matter positively electrified, which, impinging upon [the upper] plate, communicate continuously an electrical charge to the same. The opposite terminal of the condenser being connected to ground, which may be considered as a vast reservoir of negative electricity, a feeble current flows continuously into the condenser and inasmuch as the particles are ... charged to a very high potential,

1.Particle

Collector

this charging of the condenser may continue, as I have actually observed, almost indefinitely, even to the point of rupturing the dielectric[1].

This seems like a very straightforward design and would seem to fulfill his claim for having developed a fuelless generator powered by cosmic rays, but in 1900 Tesla wrote what he considered his most important article in which he describes a self-activating machine that would draw power from the ambient medium, a fuelless generator, that is different from his Radiant Energy Device. Entitled "The Problem of Increasing Human Energy - Through the Use of the Sun," it was published by his friend Robert Johnson in *The Century Illustrated Monthly Magazine* for June 1900 soon after Tesla returned from Colorado Springs where he had carried out an intensive series of experiments from June 1899 until January of 1900.

The exact title of the chapter where he discusses this device is worth giving in its entirety:

A DEPARTURE FROM KNOWN METHODS - POSSIBILITY OF A "SELF ACTING" ENGINE OR MACHINE, INANIMATE, YET CAPABLE, LIKE AN LIVING BEING, OF DERIVING ENERGY FROM THE MEDIUM - THE IDEAL WAY OF OBTAINING MOTIVE POWER

Tesla stated he first started thinking about the idea when he read a statement by Lord Kelvin who said it was impossible to build a mechanism capable of abstracting heat from the surrounding medium and to operate by that heat. As a thought experiment Tesla envisioned a very long bundle of metal rods, extending from the earth to outer space. The earth is warmer than outer space so heat would be conducted up the bars along with an electric current. Then, all that would be needed is a very long power cord to connect the two ends of the metal bars to a motor. The motor would continue running until the earth was cooled to the temperature of outer space. "This would be an inanimate engine which, to all evidence, would be cooling a portion of the medium below the temperature of the surrounding, and operating by the heat abstracted[2]," that is, it would produce energy directly from the environment without "the consumption of any material."

Tesla goes on in the article to describe how he worked on the development of such an energy device, and here it takes a bit of detective work to focus on which of his inventions he meant. He wrote that he first started thinking about deriving energy directly from the environment when he was in Paris during 1883, but that he was unable to do much with the idea for several years due to the commercial introduction of his alternating current generators and motors. It was not "until 1889 when I again took up the idea of the self-acting machine[3]."

THE TURBINE

He quickly came to realize that an ordinary electrical machine, like his generator, would not be able to directly extract energy from the cosmos and turned his efforts to what he called a "turbine" design.

The best known turbine, that is, water pump, associated with Tesla is his patent for such a device, #1,061,206, which was filed for in 1909 and granted in 1913. The unique point about this water pump is that instead of using some form of paddle wheels inside a box to move the water, he discovered that more water could be moved faster by using a set of flat metal disks. The turbine is, in itself, fascinating and may yet prove to be another important overlooked invention, but what is of concern regarding the electrical design is the general shape of the turbine - metal disks turning inside a supporting box.

This same shape turns up in another patent, this one for a "Dynamo-Electric Machine." This patent was filed and granted in the same year that Tesla said he returned to work on the "self-activating" machine, in 1889. The dynamo consists of metal disks that are rotated between magnets to produce an electric current.

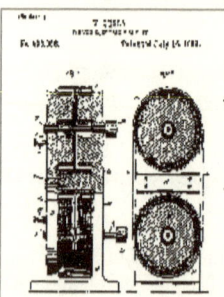

2. Dynamo Electric Machine

Compared to his alternating current generator, this "dynamo" represents something of a curious throwback to the days of Faraday's early experiments with a copper disk and a magnet. Tesla makes some improvement over the Faraday setup by using magnets that completely cover the spinning metal disks and he also adds a flange to the outside of the disks so current can be taken off more easily - all of which makes for a better generator than Faraday's. On the surface, though, it is hard to see why Tesla patented such an anachronistic machine at this point in his work.

The next piece of the puzzle is found in an article Tesla wrote for *The Electrical Engineer* in 1891 entitled "Notes on a Unipolar Dynamo." Here Tesla presents an in-depth analysis of the Faraday disk generator, explains why it was an inefficient generator, describes his improved variations on the Faraday machine, and, at the bottom of the third page of the article, states that he has devised a generator in which "the current, once started, may then be sufficient to maintain itself and even increase in strength[4]." Then, at the close of the article, he states that "several machines ... were constructed by the writer two years ago ..."[5] Two years before the writing of that article was 1889. All the evidence points to the turbine-shaped Unipolar Dynamo as being Tesla's first design for a machine that can continue to produce electricity after being disconnected from an outside source of power.

SELF-SUSTAINING CURRENT

Before going into the details of this invention it would be worthwhile to have an idea of how any generator, even in theory, could be capable of producing a self-sustaining current. This has been clearly explained by Walter M. Elsasser in a *Scientific American* article (May 1958) titled "The Earth as a Dynamo."

Elsasser models the earth-dynamo, conveniently for this explanation, on the Faraday generator of a metal disk spinning over a bar magnet placed at the edge of the disk. He notes, also, that the bar magnet could be replaced by an electromagnet which

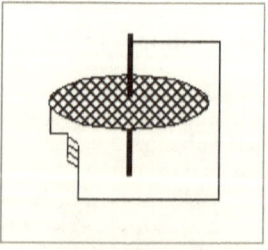

3. Faraday Generator with electromagnet

could get its power from the spinning disk by attaching one end of the electromagnet's wire to the outside of the disk and the other end of the wire to the metal rod running through the center of the disk.

Elsasser then points out that an ordinary disk generator "could not maintain a current for very long because the current induced in the disk is so weak that it would soon be dissipated by the resistance of the conductor [the disk]." This conventional arrangement would not be an answer to "how currents could be built up and perpetuated to maintain the earth's magnetic field." He does, though, propose three options in the dynamo model that would explain the earth's persistent magnetism.

If we had a material that could conduct electricity a thousand times better than copper, the system would indeed yield a self-sustaining current. We could also make it work by spinning the disk very fast... a third way we could make such a dynamo self-sustaining ... is to increase the size of the system: theory says that the bigger we make such a dynamo, the better it will function. If we could build a coil-and-disk apparatus of this kind of scale of many miles, we would have no difficulty in making the currents self-sustaining[6].

Tesla did not have a material a thousand times more conductive than copper, neither was he able to spin a disk at the ultra-high speeds needed to produce such a current, nor did he plan on using a piece of rotating metal several miles in diameter. What he did was to use energy that is usually wasted in a generator and turn it into a source of power.

UNIPOLAR DYNAMO

Tesla's design varied from that of Faraday in two major ways. First, he used a magnet that was bigger in diameter than the disk so that the magnet completely covered the disk. Second, he divided the disk into sections with spiral curves radiating out from the center to the outside edge.

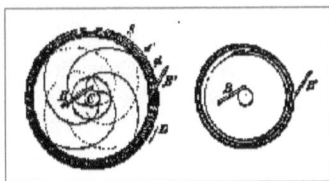

4. Tesla's Unipolar Generator

In the Faraday unipolar generator "the current," as Tesla noted, "set up will therefore not wholly pass through the external circuit ... and ... by far the greater portion of the current generated will not appear externally..."[7] By having the magnet completely cover the disk, Tesla made use of the whole disk surface in current generation instead of only a small section directly adjacent to the bar magnet, as happened in the Faraday device. This not only increases the amount of current generated, but, by making the current travel from the center to the outside edge, makes all of that current accessible to the external circuit.

More importantly, these modifications on the Faraday design eliminated one of the biggest problems in any physical system - the reaction to every action. It is this reaction that works to cancel out whatever effort goes into causing the original action. In an electrical system if there are two turns of wire wound next to each other and a current is sent through the wire, the current passing through the first loop will set up a magnetic field that will work against the current passing through the second loop.

The spiral divisions in the disk cause the current to travel the full radius of the disk or, as in his alternative version of the generator, to make a full trip around the outside edge of the disk. Because the current is flowing in a large circle at the rim of the disk, the magnetic field created by the current not only does not work against the field magnet above the circular plate, as in conventional generators, but it actually reinforces the magnet. So as the disk cuts the magnetic lines to produce a current, the current coming off of the disk strengthens the magnet, allowing it to produce even more current.

Like conventional direct current generators, the unipolar dynamo also functions as a motor if current is put into the disk while under the magnet, and this seems to be the last element that could make the device self-sustaining, that is, capable of generating a current after being disconnected from an outside source of movement like falling water or steam.

Rotation is started by, say, a motor powered by line current. Both a generator and a motor disk are mounted in the magnetic enclosure. As the disks gain speed, current is produced which, in turn, reinforces the magnets, which cause more current to be generated. That current is, likely, first directed to the motor disk which increases the speed of the system. At a certain point the speed of the two disks is great enough that the magnetic field created by the current has the strength to keep the dynamo/motor going by itself.

What process might have kept the unipolar dynamo operating after the powered start-up is speculation at this point, however two features of the generator are significant. First, when a resistive load, like a light bulb is added to the circuit, it lowers the voltage at the center of the disk. This lower voltage at the center means that there is a greater difference in voltage between the center and the outside edge of the disk than there was before the light bulb was added. As the difference between the center and the outside increases, the dynamo works harder and makes more current. Second, yet more important, the dynamo takes either very little, or no energy to keep going because the current coming off the generator is doing double duty. The current makes the bulb glow, but on its way from the generator to the filament in the bulb, it travels a path that adds to the momentum of the dynamo and, therefore, consumes energy at a very low rate. The process continues , it would seem, until heat losses in the filament equal the rotational energy of the generator's flywheel.

In terms of Elsasser's criteria for a self-sustaining generator, the Tesla unipolar dynamo comes closest to satisfying the condition of a better electrical conductor. It is not that a new material is used, but a new geometry is applied so that the current does not create its own opposing forces. This is similar, but not equivalent, to having a better conductor.

Whether or not the dynamo is in fact a "fuelless" generator it appears to be an ingenious feat of engineering that takes one of the basic principles of nature, an equal and opposite action for every action, and turns it, by the use of a novel circuit geometry, into a reaction that is additive to the original action. Instead of the opposite reaction slowing down the system that created it, the reaction adds energy to the system.

Tesla, however, was not satisfied with his mechanical self-sustaining generator. The dynamo would provide the energy to run a single machine, but his vision was to light cities and in the 1900 *Century* magazine article he elaborated on the theory of such a machine.

Imagine, he suggested, an enclosed cylinder with a small hole in it near the bottom. Let us say that this cylinder, he added, contains very little energy but that it is placed in an environment that has a lot of energy. In this case, energy would flow from the outside environment, the high energy source, through the small opening at the bottom of the cylinder, and into the cylinder where there is less energy. Also suppose that as the energy passing into the cylinder is converted into another form of energy as, for example, heat is converted into mechanical energy in a steam engine. If it were possible to artificially produce such a "sink" for the energy of the ambient medium then "we should be enabled to get at any point of the globe a continuous supply of energy, day and night[8]."

He continued, in the article, to elaborate on his energy pump but changed the image slightly. On the surface of the earth we are at a high energy level and can imagine ourselves at the bottom of a lake with the water surrounding us equal to the energy in the surrounding medium. If a "sink" for the energy is to be created in the cylinder, it is necessary to replace the water that would flow into the tank with something much lighter than water. This could be done by pumping the water out of the cylinder, but when the water flowed back in, we would only be able to perform the same amount of work with the inflowing water as we did when it was first pumped out. "Consequently nothing would be gained in this double operation of first raising the water and then letting it fall down."

Energy, though, can be converted into different forms as it passes from a higher to a lower state. He said, "assume that the water, in its passage into the tank, is converted into something else, which may be taken out of it without using any, or by using very little power[9]." For example, if the energy of the ambient medium is taken to be the water, oxygen and hydrogen making up the water are the other forms of energy into which it could change as it entered the cylinder.

Corresponding to this ideal case, all the water flowing into the tank would be decomposed into oxygen and hydrogen ...and the result would be that the water would continually flow in, and yet the tank would remain entirely empty, the gases formed escaping. We would thus produce, by expending initially a certain amount of work to create a sink for...the water to flow in, a condition enabling us to get any amount of energy without further effort[10].

Tesla recognized that no energy conversion system would be perfect, some water would always get into the tank, but "there will be less to pump out than flows in, or, in other words, less energy will be needed to maintain the initial condition than is developed [by the incoming water], and this is to say that some energy will be gained from the medium[11]."

He found that this pumping could be done with a piston "not connected to anything else, but was perfectly free to vibrate at an enormous rate[12]." This he was able to do with his "mechanical oscillator," a steam-driven engine used for producing high frequency currents. The faster the pump would work, the more efficient it would be at extracting energy from the cosmos. Research along this line culminated in the oscillator demonstrated at the Chicago World's Fair in 1893. It was not until much later, in the 1900 article, he revealed: "On that occasion I exposed the principles of the mechanical oscillator, but the original purpose of this machine is explained here for the first time [13]."

It was also in 1893 that Tesla applied for a patent on an electrical coil that is the most likely candidate for a non-mechanical successor of his energy extractor. This is his "Coil for Electro-magnets," patent #512,340. It is another curious design because, unlike an ordinary coil made by turning wire on a tube form, this one uses two wires laid next to each other on a form but with the end of the first one connected to the beginning of the second one.

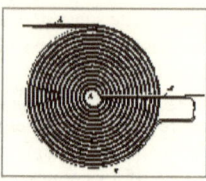

5. Coil for Electro-Magnets

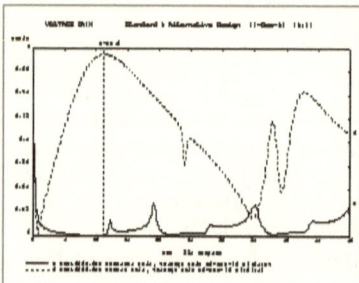

6. Voltage gain comparison

In the patent Tesla explains that this double coil will store many times the energy of a conventional coil[14]. Preliminary measurements of two helices of the same size and with the same number of turns, one with a single, the other with a bifilar winding, show differences in voltage gain[15]. In figure 6, the upper curve is from the Tesla design, the lower was produced by the single wound coil. The patent, however, gives no hint of what might have been its more unusual capability.

In the *Century* article Tesla compares extracting energy from the environment to the work of other scientists who were, at that time, learning to condense atmospheric gases into liquids. In particular he cited the work of a Dr. Karl Linde who had discovered what Tesla described as a "self-cooling" method for liquefying air. As Tesla said, "This was the only experimental proof which I was still wanting that energy was obtainable from the medium in the manner contemplated by me[16]."

What ties the Linde work with Tesla's electromagnet coil is that both of them used a double path for the material they were working with. Linde had a compressor to pump the air to a high pressure, let the pressure fall as it traveled through a tube, and then used that cooled air to reduce the temperature of the incoming air by having it travel back up the first tube through a second tube enclosing the first[17]. The already cooled air added to the cooling process of the machine and quickly condensed the gases to a liquid.

Tesla's intent was to condense the energy trapped between the earth and its upper atmosphere and to turn it into an electric current. He pictured the sun as an immense ball of electricity, positively charged with a potential of some 200 billion volts. The earth, on the other hand, is charged with negative electricity. The tremendous electrical force between these two bodies

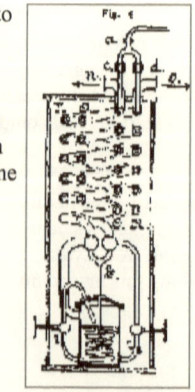

7. Linde's condenser

constituted, at least in part, what he called cosmic energy. It varied from night to day and from season to season but it is always present.

The positive particles are stopped at the ionosphere and between it and the negative charges in the ground, a distance of 60 miles, there is a large difference of voltage - something on the order of 360,000 volts. With the gases of the atmosphere acting as an insulator between these two opposite stores of electrical charges, the region between the ground and the edge of space traps a great deal of energy. Despite the large size of the planet, it is electrically like a capacitor which keeps positive and negative charges apart by using a non-conducting material as an insulator.

The earth has a charge of 90,000 coulombs. With a potential of 360,000 volts, the earth constitutes a capacitor of .25 farads (farads = coulombs/volts)[18]. If the formula for calculating the energy stored in a capacitor ($E = 1/2CV^2$) is applied to the earth, it turns out that the ambient medium contains 1.6×10^{11} joules or 4.5 megawatt-hours of electrical energy.

In order to tap this energy storehouse Tesla had to accomplish two things - make a "cold sink" in the ambient energy and devise a way of making the "sink" self-pumping. Explaining how this process might have worked requires, again, speculation.

Such a "sink" would have to be at a lower energy state than the surrounding medium and, for the energy to continually flow into it, the "sink" would have to maintain the lower energy state while meeting the power requirements of the load attached to it. Electrical energy, watt-seconds, is a product of volts x amps x seconds. Because the period of oscillation does not change, either voltage or current has to be the variable in the coil's energy equation.

In that the double wound coil maximizes the voltage difference between its turns, it is probable that it is the current that is minimized to produce a low energy state in the coil. For the coil to be initially "empty" and at low energy would mean it operated at high voltage with a small amount of charge[19].

The coil, then, would be set into oscillation at its resonant frequency by an external power source. During a portion of its cycle the coil will appear to the earth's electric field as one plate of a capacitor. As the voltage across the coil increases, the amount of charge it can "sink" from the earth's higher energy field will increase.

The energy taken into the coil - through the "small opening" which appears to be the atomic structure of the conductor according to the physics of Tesla's time - is "condensed" into positive and negative components of current, a lower energy state relative to the originating field.

The current is equivalent to the water converted to gases in Tesla's description of the self-acting engine. The current would "escape" from the "sink" into whatever load was connected across the coil. The movement of current into the load would produce a strong magnetic field (the stated intention of the patent) which, when it collapsed, would, again, produce a high potential, low charge "sink" to couple with the earth's electric field.

Because the inflowing energy performs a double function similar to the unipolar generator, supplying current to the load and aiding the pumping function, the system's energy expenditure in moving charge is low, allowing the system to gain more energy from the medium than it expends in its operation. The coil needs no extra energy from an outside source to pump the energy it has extracted.

Energy would come directly from the sun.

A more modern view of such a device, should it prove to operate in this theoretical manner, would be to describe it as a self-oscillating capacitive system. Once the device is set into oscillation, very little power is expended in driving the load. Because it is an electrostatic oscillating system, only a small amount of charge moves through the load per cycle, that is, the coulomb per seconds = amps are low. If the charge is used at a low rate, the energy stored in the capacitive system will be turned into heat at a slow rate enabling the oscillations to continue for a long period of time.

With his prominent position in the world of science at the time, it is curious why Tesla's invention was not commercialized or at least publicized more. Economics, not science, appear to have been the main factor. The adoption of alternating current was opposed by powerful financiers of the period. Michael Pupin, another leading electrical researcher at the turn of the century, noted in his autobiography:

...captains of industry...were afraid that they would have to scrap some of their direct current apparatus and the plants for manufacturing it, if the alternating current system received any support ... ignorance and false notions prevailed in the early nineties, because the captains of industry paid small attention to highly trained scientists [20] [21].

Tesla's patents for electrical generators and motors were granted in the late 1880's. During the 1890's the large electric power industry, in the form of Westinghouse and General Electric, came into being. With tens of millions of dollars invested in plants and equipment, the industry was not about to abandon a very profitable ten year old technology for yet another new one.

Tesla saw that profits could be made from the self-acting generator, but somewhere along the line he had pointed out to him the negative impact the device would have. At the end of the section in *Century* where he described his new generator he wrote:

I worked for a long time fully convinced that the practical realization of the method of obtaining energy from the sun would be of incalculable industrial value, but the continued study of the subject revealed the fact that while it will be commercially profitable if my expectations are well founded, it will not be so to an extraordinary degree [22].

Years later, in 1933, he was more pointed in his remarks about the introduction of his fuelless generator. In the Philadelphia *Public Ledger* of November 2nd, is an interview with Tesla under the headline "Tesla 'Harnesses' Cosmic Energy." In it he was "Asked whether the sudden introduction of his principle would upset the present economic system, Dr. Tesla replied, 'It is badly upset already.' He added that now as never before was the time ripe for the development of new resources."

It has been nearly a century since Nikola Tesla claimed a radically new method for producing electricity. The need for the development of new resources is greater now than at the end of the last century. Perhaps these overlooked inventions will make his vision of "increasing human energy through the use of the sun's energy" become a reality.

In 1884, at the dawn of the electrical age, a slender, well-dressed young man stepped off a ship in New York with only four cents in his pocket. But in his mind he carried inventions that would revolutionize the world.

The man was Nikola Tesla. He was born in Croatia, in southeast Europe, in 1856, and when he was still a little boy he became fascinated by electricity. On winter nights he would pet his cat and watch with wonder as "my hand produced a shower of crackling sparks."

Lots of other things fascinated young Nikola. To learn how clocks worked, he took them apart ("I was always successful," he said), then tried to rebuild them ("at this, I often failed"). He built waterwheels and other machines, including a motor powered by 16 June bugs glued to a spinning disk. (He was proud of this invention and showed it to a friend, who promptly ate some of the bugs. Nikola threw up.) Watching the pigeons that flew around his family's small farm made Nikola curious about flight. He climbed to the roof of a barn, breathed in and out rapidly to make himself feel lighter, then jumped into the breeze holding an umbrella. He spent several days in bed recovering from the landing.

As a child, young Tesla spent a lot of time in bed because he was often sick. But during those long weeks he created magnificent inventions in his mind. He was an excellent student, and when he was 19 he left home to study engineering.

One of his professors showed Tesla an invention that could be connected to an electrical generator to act as a simple motor. Tesla studied it, then told his professor it would be a much more efficient motor if it were powered by alternating current. (Electricity comes in two flavors. Direct current -- DC -- flows through a wire in one direction only, while alternating current -- AC -- changes direction constantly, flowing forwards and backwards). But no one knew how to build an AC motor, and Tesla's professor told him it was impossible.

In 1881, Tesla left the university. He got sick again. His doctors called it a nervous breakdown (we'd probably call it depression). When he finally got better, Tesla realized his mind had been working on the AC motor the whole time -- and one day, walking with a friend, he had the answer! "The idea came like a flash of lightning," he said, "and in an instant the truth was revealed." Tesla picked up a twig and began drawing in the dirt. "See my motor," he said to his friend. "Here."

Soon afterward, Tesla stepped off that boat in New York, where Thomas Edison was building generators and developing an electrical system. Edison used direct current, which couldn't be transmitted more than a mile. It worked too inefficiently. That meant that even a small town would need hundreds of power stations. Tesla knew his AC system was better--its current could be transmitted hundreds of miles. Edison, however, was stubborn. He told Tesla he wasn't interested.

But Edison had a rival in the race to develop electricity. His name was George Westinghouse, and he was very interested in alternating current. He bought Tesla's ideas, and developed an AC system that lit up the 1893 Chicago World's Fair. Three years later an AC system began turning the energy of Niagara Falls into electricity.

Tesla's invention, says Barney Finn, curator of electricity at the National Museum of American History, "gave life to alternating current," and ended Edison's dream of electrifying America with direct current. The motors we use every day--all of them--to run electric drills, to open cans of cat food, or to power refrigerators and hair dryers, are based on the idea that came to Tesla like a flash of lightning.

Soon Tesla, who proudly became an American citizen in 1891, was one of the most famous inventors in the world. He loved to entertain friends like Mark Twain with hair-raising demonstrations of AC. Dressed in a tuxedo, he would stand near a huge generator while a million volts of high-frequency current sent lightning dancing over his body. In his hands, light bulbs shone like suns and gas-filled tubes blazed like Darth Vader's light saber.

Tesla didn't stop experimenting. He found that high-frequency alternating current, transformed by the device that's still called a "Tesla coil," could send signals through the air without wires; he had invented radio -- the "wireless."

Tesla built a laboratory in Colorado Springs to study electricity. He created lightning bolts that flashed more than 100 feet into the Rocky Mountain night. The equipment--a 180-foot metal tower--was so powerful that it set fire to the Colorado Springs power plant. Then Tesla decided to build an even more powerful tower near New York. He hoped it would beam electricity around the world--but he couldn't raise the money for the project.

His fame began to flicker out. Edison--who had fought Tesla's AC system--had become known as the wizard of electricity; Guglielmo Marconi was called "the father of radio," although the Supreme Court took away some of Marconi's patents because Tesla (and others) had come up with the ideas first.

Why did the world remember men like Edison yet forget Nikola Tesla? For one thing, Tesla was a terrible businessman. He didn't patent his work enough; he talked himself out of a good contract with Westinghouse for his AC system; he couldn't drum up curiosity using the media the way Edison did.

Tesla was nearly forgotten when he died in 1943. His only friends seemed to be the sick pigeons he brought home and nursed back to health in his New York apartment. But he always believed that "harnessing the forces of nature to meet human needs" was the most important contribution one could make. "I have had more than my full measure of this exquisite enjoyment," Nikola Tesla said. "My life was little short of continuous rapture."

The Tesla Electro-Magnetic Motor

Born in Croatia, the engineer Nikola Tesla had a distinguished academic and industrial career in central and eastern Europe before coming to the United States in 1884. Here, while working for the Edison Machine Works and independently, Tesla created his greatest invention, the electro-magnetic motor.

A motor converts electrical energy to mechanical power by using current to make a metallic loop (the "rotor," or "armature") spin around a central shaft. Tesla was convinced that DC ("direct current") motors could be modified to operate without commutators---external switches that reverse the direction of the current in the rotor every 180 degrees to keep it spinning in one direction. In early 1888, working out of his experimental shop in New York, he proved his theory: Tesla built and demonstrated the "induction" or "electro-magnetic" motor (patent #381,968). Tesla's revolutionary motor used a rotating magnetic field, rather than mechanical switches, to spin the rotor. This made unit drives for machines possible, and allowed the more efficient AC power ("alternating current," where the rotor swings back and forth) to become the standard for most office and household appliances. Tesla sold his invention to the recently-founded Westinghouse Electric Company, which might not have flourished without his contributions.

Tesla was also a pioneer in the early days of radio (invented by Guglielmo Marconi at the turn of the century). In the year of Tesla's death (1943), the U.S. Supreme Court ruled that Tesla's patents for the radio superseded those of Marconi: this makes Tesla the father of the second generation of radio.

By the end of his career, Tesla had over 700 inventions and 100 patents to his credit. Though his innovations never made him wealthy, Tesla is rightly renowned to this day as one of the greatest electrical engineers of all time.

.

Welcome to the world where free energy abounds providing unlimited clean energy 24/7.

It's a here and now thing.

Welcome to our world. Free energy, it was also a then thing.

Were you aware that Free energy has proved itself as a real technology for 157 years now?

Alexander Graham Bell was the first in history to prove the reality of free energy in 1855.

Welcome to the universe. We've been invited to live there in peace with our elders. Welcome.

How many of us have heard the claim that the powers that be have been blocking us from experiencing the

best that technology has to offer us? Could there be any truth to this? Hopefully this book will answer that question for you.

2

Read of the work of some of the experimental researchers that have been looking to help us ride the free

energy express. We'll visit the workshop of Michael Faraday (1791-1867). Revisit his early discovery of

the nature of the specific differences between the two components, electrical and magnetic, that comprise

electricity.

How these 2 components of electricity exhibit particular properties and act in distinct ways that can make

them useful to extract free energy from the universe.

Nathan Stubblefield (1860-1948). What did he discover about electricity that allowed him to keep his home

toasty warm through the coldest days of winter in the early 20th century using free energy that didn't cost

him a penny?

Dr. Henry Moray (1892-1974) had a solid-state transistorized free energy machine up and running in 1928

that daily produced $120 dollars worth of electricity at today's prices. When asked, "Where's the energy

coming from." He responded: "I don't know, but I think it has something to do with particle size."

Edwin Gray (1925-1989) Built a cold running electric motor that puzzled the top scientists of the time:

"Why did it run so cool? What does cold electricity have to do with free energy?"

The exciting world of free energy. Here and now and then. Velocity power sources. Machines for a world

full of people all made to live together in peace.

Nicola Tesla (1856-1943) had given a proposal to JP Morgan in 1906 to build a 5,000 light bulb, free

energy lighting system for illuminating a sports arena at night. JP and associates declined the opportunity to

light the way for humanity into a better day. He pulled down the laboratory Tesla was using for his

research, JP also cut off Tesla's bourse. How painful was that?

The power of bourse: "Is that really all the power there is?"

Can we only wonder how different things would be today if not for the decision to deny the human race the

best that our science and labor could build in 1906? What would the world be like today if only they had

allowed this free energy technology in? No doubt its best to let go of the past, so can we think forward of

What joy there will be when we build and install this life saving technology?

Do we have some catching up to do here? Is it true that if there's a will there's a way?

Any doubt that we have the will to build the best machines to feed, house and clothe all of God's kids?

Can we imagine in how many wondrous ways free energy will change the circumstances of life for the

mass of the human race?

When we understand that our planet is in high-speed motion all of the time will we finally have our clues

that will let us connect the dots and build as many free energy machines as we want?

The history of the search for free energy extends in the modern era from the time of Michael Faraday and

his laboratory experiments in the 1830s' until current events of today. The hex controller is one of the

circuits we will take a look at in this book.

Designed and built in California, the hex controller has-been charging batteries for three years now using

free energy. There is a link to a site with a video showing an experimenter doing just that; Charging his

Childs electric tricycle using only the hex controller.

3

If seeing is believing, After you see it with your own eyes, is it likely you'll be a believer?

End forever any doubt whatsoever that free energy is a here and now thing?

The hex controller is now on the shelves and waiting for the builders to use it, or one like it. It is the real

thing. For over a century now many other free energy researcher builders have been creating many devices

to make lives better. Working to make a world where hope keeps shining on everyone. Will you be helping

to build them?

We'll take a look at the properties of electricity and some of the researchers and their machines and what it

is they've found that can be so useful to us as we un-puzzle the mystery of free energy and where its

coming from. Free energy, Velocity power sources; making it understandable.

Do we have the labor available that is ready willing and able to build the machines that will help enable us

to live in a world of peace and plenty for all?

Will the certainty and proof of free energy help to finally bring all hands on deck to aid in placing the rule

of peace into its proper place; our present and future?

Peace be with you. Does that sound real nice to begin?

Wind power is promoted as clean energy. It's nice when its up and running. When the Wind stops blowing,

what then?

With the free energy derived from planetary motion on line, power is available all of the time; 24/7 from

now until eternity.

One thing you can be sure of, Free energy created with the use of over unity devices is not a make believe

technology; It's the real thing.

The thing that has been missing until now is the where and how to explain what has been continuously

demonstrated by numerous engineers and scientists over the many years of the history of free energy

machines; a useful and good theory of operation.

The refrain from the choir of skeptics is that free energy cannot be real because it violates the first and

second laws of thermodynamics. Basically what they say is that we can't have more energy coming out of a

machine than what is going into it.

How long could it have taken our early ancestors to put it together that when the wind was blowing, a sail

on their boat would help them on their journey?

With the wind we can feel it. Free energy is invisible as to where it is coming from, and besides that it took

thousands of years to puzzle it out that our planet is in motion to begin with.

That's the key to understanding it all. High speed Planetary motion. That's what allows for extraction of

free energy.

A theoretical explanation of where free energy is coming from is why this book has been written, with a

prayer and a hope to inspire those who will build these wondrous machines of life and peace to go faster.

"War? What war?. We don't need no war!" War provides advantages we just don't need.

4

Machines that produce over unity energy are a well-proven fact of life.

We'll start in the good old days of free energy, the first scientists of the modern age that were looking for

methods to reliably tap into the "Power of the Cosmos." Michael Faraday was the first to search for it in

planetary motion. Dr. Mallove looked at shearing force oil heaters and Cold fusion in our modern age for it.

Dr. Eugene Mallove (1947-2004) said something to the effect shortly before he left us that: "We're only a

few months away from clean free unlimited heat and electrical energy production."

Little has changed since Dr. Mallove left us. Be assured, we're only a few months away from unlimited

production of clean free heat and electrical energy. Control of gravity is also part of the new science of free

energy technology. Townsend Brown (1905-1985), discovered gravity power in 1921 using high voltages

and a transformer to produce it. It's all a dimensional thing. We'll explain that to.

Please read on my friends.

NOTE: This is a composite book. I have used several things found on the Internet and put them in here as I

found them, some I rewrote to connect things together. It's the theoretical understanding of the technology that is from my hand. So for those who have put their writing in to this, I thank them. When using other

peoples writing hopefully there should be a link to the entire piece they wrote.

· Chapter One

· Michael Faraday

What he knew and when he knew it?

Michael Faraday, FRS (22 September 1791 – 25 August 1867)

Was a British scientist, chemist, physicist and philosopher who greatly contributed to the fields of

electromagnetism and electrochemistry. His main discoveries include that of the Magnetic Field,

Electromagnetic Induction, Diamagnetism and Electrolysis.

Although Faraday received little formal education and knew little of higher mathematics such as calculus,

he was one of the most influential scientists in history;[1] historians[2] of science refer to him as having

been the best experimentalist in the history of science.

It was by his research on the magnetic field around a conductor carrying a DC electric current that Faraday

established the basis for the concept of the electromagnetic field in physics.

Faraday also established that magnetism could affect rays of light and that there was an underlying

relationship between the two phenomena.

He similarly discovered the principle of electromagnetic induction, diamagnetism, and the laws of

electrolysis. His inventions of electromagnetic rotary devices formed the foundation of electric motor technology, and it was largely due to his efforts that electricity became viable for use in technology.

http://en.wikipedia.org/wiki/Michael_Faraday

While Faraday was doing his electrical experiments he made several attempts to tap into planetary motion.

Using several hundred feet of iron and copper wires he ran them across the lawn and put the ends into a

5

pond of water. He at first thought he detected a response, but soon later decided that it was anomalous

factors. He felt that he did not detect the presence of free energy.

Is it a curiosity that he specifically spoke of tapping into planetary motion? He was quite aware at the time

that our planet earth is rotating around our sun at 66,000 miles per hour. That translates into 18.5 miles per

second. He is the first researcher this writer has uncovered to specifically think and speak of tapping into

the high-speed motion of our planet.

Faraday made many discoveries, one of which was the discovery of two static fields of planet earth;

possibly four?

Reading his notes requires a translation often. Not from a different language, rather a different time. He

was finding things out that were so new we have to take time to translate to make understandable what it

was that he actually found. The word electron entered our language in relation to electricity in 1891, many

years after Faraday's time.

From looking at his notes though it seems he discovered at least 2 static fields.

About 80 years later, in about 1910-1911 Tesla discovered 3 static fields.

As a thunderstorm approached his laboratory Tesla noted that a lightening strike nearby caused 3 of his

meters that he had attached to metal coils in his lab, jumped to life; and stayed that way for an extended

period of time.

After the storm had passed, Tesla noted that the electric fields that had caused his meters to respond, took

time to dissipate the electricity from the thunderstorm's activity. He deduced that the invisible static fields

of our planet were acting not unlike an electrical capacitor, storing static electricity.

He further noted that one signal was very slight, one was jittery and one was like a brick wall.

Could Tesla have at that early time detected the primary inertial field of planet earth?

These days it's reported that there are 12 planetary inertial fields.

Could some of this research information be sitting on a shelf somewhere unknown to us concerning some

of these facts of life? Has the twelve static fields been completely understood and measured by the many

space probes launched in our time? Do inquiring minds want to know?

If you have the time to search out and read more about some of the free energy researchers and builders,

you may discover how much fun is waiting there for us. Are we ready to puzzle through the questions that

free energy presents to us?

If you yourself have some knowledge or happen to come across something useful, there are a couple of

email addresses at the end of this book. Will be happy to share your contributing knowledge with others for sure.

When we have the free energy in hand is there a feeling that this will help us to rise to a higher order?

Maybe an even better question is how long will it take us to build them and bring them on line? Will we

find the answer in who controls the bourse? Will understanding free energy reveal to us how important is

the question of the bourse and who is authorized to issue it?

Michael Faraday was aware of the motion of planet earth. He stated that that was where the free energy was

to be found. Quite surprising considering that this was in the 1830s.

6

Did he somehow intuitively surmise that planetary motion could be harnessed in some way to produce

energy?

Faraday also made a discovery concerning the dual nature of electricity. He found that heating a wire that

had amperage and voltage in it would cause the voltage to disappear. The amperage in the wire remained.

This is significant in so many ways; this was the principle that allowed for the vacuum tube to make its

appearance. There is a heater in a vacuum tube that causes a separation of the electrical field from the

magnetic field, using heat. This was a key development that allowed for rapid development in the early

days of radio.

Lee De Forest's 1906 "audion" was developed as a radio detector, and soon led to the development of the

triode tube.

How many of you remember televisions that had tubes in them?

This may help give a clue as to how to separate the electro-magnetic spectrum that we call electricity.

Faraday demonstrated the dual nature of electricity. Proved that its actually 2 things. Magnetic and

electrical; each side a part of the electro-magnetic spectrum.

If you have a tube set around, you may be able to produce some free energy with it. Tesla used 12 ordinary

vacuum tubes to power a car with a 40 horsepower motor in the 1930s.

Along with laboratory experiments, Faraday was looking deeply into the theoretical aspects of electricity.

There is an issue called Faraday's paradox, questions that still are with us today. It concerns spinning a disc

with a magnet held near to it and producing the flow of electricity. When the magnet is spun around the

disc, no electricity is produced. When the magnet and the disc are spun together an electrical current is

produced.

As this paradox is explained by several varying theories, would like to add this one to it. The standing

waves of the planet are involved with the magnetic field causing the electrical flow. Here is a link if you

care to take a look at the Faraday paradox.

http://en.wikipedia.org/wiki/Faraday_Paradox

Is it a curiosity that the work of Michael Faraday in the first part of the 19th century is still being puzzled

out to this day?

There is also the work of Henry Andrews Bumstead, (1870-1920). He had focused much of his efforts for several decades looking into aspects concerning the electrostatic field.

Bumstead obtained his PhD in 1897, submitting as thesis a paper entitled A Comparison of Electrodynamic

Theories.

Unfortunately it does not seem to have been published, the only copy in existence being the manuscript in

the author's own handwriting, which has been preserved in the archives of the Yale library.

http://en.wikipedia.org/wiki/Henry_Andrews_Bumstead

AS Henry had received a good formal education, could we guess that he would have had knowledge of

planetary motion? Though he has not previously been mentioned as having anything to do with free energy

research, Can we only wonder if he put elements of planetary motion into his thinking and combined with

7

the static fields that he had spent several decades studying, drawn some early conclusions as to where the

free energy may be coming from?

American researchers concluded from cosmic particle time variations experiments done in 1941 that

relativity was a proven fact of life. Arthur Eddington (1882-1944) and other European researchers had

considered it likely that the theory of relativity was a fact of life as early as 1915.

IN 1892 Henry Andrew Bumstead (1870-1920) attended lectures by Henry Rowland concerning the

electrostatic field.

http://en.wikipedia.org/wiki/Henry_Augustus_Rowland

Previously, In 1871 Henry Rowland (1848-1901) visited with a Mr. Paine who had built what was

described as a magnetic motor that powered a saw.

http://archive.org/details/physicalpapersof00rowlrich (page 30-34)

The machine had magnets encased in thick cast iron covers with the magnets themselves wrapped in a

metallic foil that Mr. Paine invented.

Henry Rowland didn't understand how the machine could work, yet he reported it maintained absolute

accuracy of speed whether or not the saw was cutting deep into wood or only slightly. In one demonstration

the drive belt flew off the machine and the motor stayed at the exact speed. No variations?

The inventor, Mr. Paine gave technical descriptions to Henry that didn't make sense and were outside the

ordinary understanding of the technology of the time. Could this be another early inventor that had great

intuitive ability to build an over unity free energy machine, yet didn't have the theoretical knowledge to

explain what was actually driving the machine?

To share a thought, have you ever seen the demonstration of dropping a magnet through an aluminum pipe?

Here's a link to a magnet dropped in an aluminum pipe and another with a copper pipe.

http://www.youtube.com/watch?v=FocmCljrSj0

http://www.youtube.com/watch?NR=1&feature=endscreen&v=E97CYWlALEs

If you've watched the videos, is it interesting with the aluminum and copper, that the speed of the falling

magnet itself is changed when the magnet is dropped inside of the metal tubes?

Copper is diamagnetic, it will create a magnetic field in opposition to an externally applied magnetic field.

Aluminum is a paramagnetic material and is only attracted when in the presence of an externally applied

magnetic field. It does not retain magnetism when the field is removed.

Now can we think about Mr. Paine's machine in 1871 and wonder what type of metallic foil he wrapped

his magnets in? How did these interact with our planet's static fields?

Can you imagine that as with the magnet falling at a slowed rate of speed through aluminum or a copper

pipe, that the metallic foil of Mr. Paine in 1871 may have changed the field properties of the magnet?

Will this puzzle from 1871 be another clue for builders of these velocity power source machines? That

sure is the hope here.

As we cannot easily sense the motion of our planet and it took literally thousands of years to prove that it is

in motion, is it understandable that inventor's like Mr. Paine who built the free energy machines never

guessed where the energy was coming from?

8

Did you make a note of the absolute constant speed of Mr. Paines saw even when its belt flew off?

Henry Rowland had mentioned that ordinary electricity would have shown a change of some degree of

speed variation? The saw remained at the same speed in all modes of operation?

Only one other researcher besides Faraday that I've found mentioned planetary motion as a possible source

of free energy.

That man was Lester Hendershot (1899-1960) who spoke of the rotation of the planet as the source of the

motive power that was running his free energy motor. He theorized that a magnetic wind was blowing from

the earth rotation motion. He claimed to wind his coils so as to intersect with this magnetic wind.

"Invention Result Of Dream"

New York Times (Sunday, February 26, 1928)

"Fuelless Motor Impresses Experts"

Hendershot Made First Fuelless Motor For His Son's Toy Plane ~

The invention of the fuelless motor, tested at Detroit, was the result of a dream by its inventor, Lester

Jennings Hendershot, who lives on "the street back of the railroad" in this town of about 3,000 inhabitants,

15 miles from Pittsburgh.

Although Hendershot was on his way from Selfridge Field today and is not expected home until tomorrow,

his wife told of his conception of the machine and how the miniature model was constructed from the parts

of a worn out radio which had been given to the inventor by his uncle.

Several years ago the vision of a machine which would operate from "earth currents" came to Hendershot

in a dream, according to his wife, but it was not until last November that he actually started working on it.

His 4-year old boy had built a small airplane at that time and was considerably chagrined because it would

not operate. The father was disturbed too, so he told his son he would build an airplane which would work.

The result of that was the fuelless machine.

First Model Worked Toy Plane ~

When the miniature motor model had been constructed, Hendershot built a small airplane and placed the

machine in it. A switch was turned and immediately the propeller began to move. The machie was not

connected to any electrical current, but was running on is own accord from "earth currents".

For several weeks the little motor and the airplane rested upon a small table in the living room of the

Hendershot home, which faces an unpaved street near the railroad tracks. One day D. Barr Peat of Bettis

Field, the air mail port near McKeesport, Pa., visited the Hendershot home to see the model.

He immediately became enthusiastic and a few weeks later he and Hendershot were at Selfridge Field

where permission was been granted to build a model large enough to operate an airplane.

Hendershot, who is only 29 years old, was born in Hyndmann, Pa. His schooling has not been extensive,

although he spent a few months several years ago at Cornell University, where he took a few courses in

mechanics. He has not been employed at any particular task and has been known as a "freelance" worker.

He has been a fireman and an engineer on the railroad, has worked in the mills near Pittsburgh, has

inspected concrete and done electrical work. During the war he was a bugler with a machine gun company,

but did not get overseas.

9

Still Wants To See "How They Work" ~

According to his mother, he has always been interested in mechanics and when a child he would insist upon

taking his playthings apart.

And that desire has not escaped him a man, for even now he takes his own son's playthings apart to "see

how they work".

It required only a few weeks for him to construct the miniature model of his fuelless motor, although he

worked day and night during that time. He had a crude workbench in the cellar of his home, which was

placed near the furnace, where it was warm. Early in the morning he would be there, tinkering about, and

late at night he still could be found there.

Hendershot's idea was that the earth currents which make the aurora borealis in the skies could be

harnessed by man and made to produce power that would operate an engine.

The youthful inventor has no other inventions to his credit.

http://www.rexresearch.com/feg/feg1.htm

Hendershot had mentioned rotation of earth. He felt that the way he wound his coils caused the earth's

electromagnetic 'wind' to blow against them producing the motion that he demonstrated.

Both he and Michael Faraday spoke of planetary motion as to where the excess over unity free energy was

coming from. Faraday specified the 66,000 miles per hour earth around our sun rotation to where he was

looking.

Hendershot seemed to be thinking of earth axis rotation motion as the proximate source of the free energy.

Both are correct, as our planet revolves on its axis about 1,000 miles per hour.

That gives us our 24-hour rotation equaling 24,000 miles circumference of earth, our day and night by this

constant rotation of earth on its axis. Actual earth circumference is 24,903 miles.

Along with orbital Rotation around our sun, we also have a solar and a galactic system rotational motion to

add in to the picture when considering how fast we are traveling on our journey through space.

These are all factors in powering free energy machines that extract energy from the motion of earth.

Combined, earth, solar and galactic orbital speeds approach nearly 600 miles per second of total speed as

we travel through the Cosmos. The power of the Cosmos. The power we need. The power source of our

free energy. Could we consider that it is the universe itself that is available for us to tap into for all of our

energy needs?

This is mentioned because as we journey from our planet, the free energy remains available in deep space.

There are many things to learn. How many discoveries yet to be made? Are we only scratching the surface

when speaking of our planet as a power source? Is it still true that the longest journey begins with the first

step?

Here for your kind consideration is another simple device that powered a home and car nearly a century

ago.

10

Fate Magazine, October 1956, pp. 123-125; "Report from the Readers"

Mysterious Invention

"The Hubbard Energy Transformer" brought back to me exciting memories of another inventor. In 1918,

while doing painting and decorating, I was hired to paper several bedrooms in a large two-story house.

While at this work I went down to the back porch to pick up some materials. I happened to glance at the

light meter and saw it was not moving.

I opened the fuse box and saw the main power fuses had been removed. It took only a minute to make sure

the line had not been taped beyond the meter.

The only member of the family at home at the time was a young man in his early twenties [C. Earl

Ammann. I asked him, "Earl, where do you get your juice? I noticed it does not come from the power

lines".

"Come along and I'll show you", he said. He led me up to the attic. He placed some steel bars on a work

table and picked up a coil which looked like a loose coupler. After placing the coils on the steel rods he

touched the opposite terminal. The bell rang with great force, and there was quite a spark, too.

I picked up the coils to make sure there was no contact with other appliances. I could see right through

them. There was no battery inside. The bell rang just as vigorously. The wire was iron.

In the basement Earl had what he called an Activator Transformer, the size of two fists, which had to be

within 10 miles of the radius of the generator coils. The activator was not in contact with any visible wires or appliances. It was activated by the electric currents which surge around the earth and activate the

compass needle. By cutting into these currents, earl said, we can obtain unlimited power.

A year later Earl demonstrated his Cosmo Electric Generator in Denver. He had placed two copper spheres

on the front fenders of his car in pace of the headlights. From these copper spheres he obtained enough

power to drive that old jalopy all over Denver as reported in the Denver Post at the time.

While Earl was demonstrating his invention all over the streets of Denver, the power had been cut off in the

foothills. In spite of this, when he went to Washington DC shortly afterward to try to obtain a patent on his

Cosmo Electric Generator, he found that charges had been filed against him claiming he had a device to

steal power from the power lines.

K. H. Isselstein,

Spokane, WA

http://www.rexresearch.com/feg/feg1.htm

"Activated by the electric currents which surge around the earth and activate the compass needle. By

cutting into these currents, earl said, we can obtain unlimited power."

Here again we have an inventor who is thinking his free energy is coming from the fields of planet earth,

and Intuitively he is able to build a machine to extract this energy.

No theoretical understanding that can explain it though?

Are we long since past the time that the free energy is now well enough understood that soon it will be a

certain promise for us?

One of the most interesting researchers and builders in the early times was Nathan Stubblefield. Have you

ever felt that these things appear to be so far ahead of us even today that it seems impossible that scientists

of long ago could have really demonstrated free energy? Please Read on About Nathan and the wonder of

his world. Heated his home toasty warm with free energy. Why do so many remain out in the cold? Didn't

Nathan solve that for us long ago?

· Chapter Two

· Nathan Stubblefield

Heaters, motors and light. Nathan had it all as early as the 1890s.

From Wiki

Nathan B. Stubblefield (November 22, 1860 - March 28, 1928) was an American inventor and Kentucky

melon farmer.

It has been claimed that Stubblefield invented the radio before either Nikola Tesla or Guglielmo Marconi,

but his devices seem to have worked by audio frequency induction or, later, audio frequency earth

conduction [1] (creating disturbances in the near-field region) rather than by radio frequency radiation for

radio transmission telecommunications.

On January 1st 1902, (though one source mentions a date 10 years eariler) he publicly demonstrated voice

and music transmissions to five receiving locations on the courthouse square in Murray witnessed by at least 1,000 persons, apparently using voice frequency transmission through earth conduction, to a radius of

a half a mile.

Later on March 20th 1902, he demonstrated wireless telephony in Washington D.C. where voice and music

transmissions were made over a third of a mile from the steamer Bartholdi to shore.

He also demonstrated wireless telephony in Philadelphia where, on May 30th 1902, a distance of a half

mile was, again, achieved.

Tests in New York City in June the same year were less successful because of electrical interference from

the alternating current power in widespread use in the city.

Stubblefield knew that rods in the ground, placed into various spots, reveal an amazing degree of electrical

activity.

The currents varied across plots of ground. Wet soils often reverse the expected electrical strength;

weakening rather than strengthening their current.

Stubblefield knew that a proper placement of metallic ground probes could produce stronger currents for

him to use, but he did not anticipate what he then accidentally discovered.

His initial experiments involved the development and examination of simple earth batteries: buried metallic

arrangements, which produced weak electrolytic power, during the early stages of this charge building

process, the characteristic weak output was observed.

This was usually a volt at half an amp, the general electrolytic output of buried metals.

However, if properly placed, the energetic output of his cell would be phenomenal.

Finding such a power point, he buried one of his cells. The process took a week or more to build strength.

12

Once the cell was "saturated" it became (in his words) "a conduit of earth charge".

Stubblefield simply stated that the fully saturated coil suddenly "manifested an electromotive force far

greater than any known wet-cell".

This state being achieved, the current flowed in "commercial electrical volumes".

Stubblefield developed a peculiar bi-metallic induction coil which, when buried, draw up sufficient

electrical power to operate lamps and other appliances which he designed and tested.

A great length of both cotton-insulated copper and bare iron wires were wound together in a "bifilar"

arrangement on a large iron stove bolt.

The windings were held side by side throughout the coil.

His patent specification describes the device as a "terminal, which draws electricity out of the ground".

Where did this extra energy come from? From what mysterious depths did this strange power emerge?

Was it electricity as we know it? It has been suggested that earth energy, the pre-electrical energy of the

ground, was at work in all these systems.

Called "vital energy" by Victorian Science, this presence exceeded the character and nature of ordinary

electricity.

AS a note, iron and copper that Nathan used have distinct properties in relation to magnetism. Iron is magnetic, copper is diamagnetic. The free energy saw Mr. Paine demonstrated had metallic foil shielded

magnets in an Iron container. Copper or aluminum metallic foil possibly?

Electrical power grew to spark potentials in these lines when no exterior evidence allowed explanation of

the energy levels.

Some of Nathan's ground batteries that should have operated for a few months at most, continued to

provide electricity for several years until the telegraphs they powered were no longer in use. When the

batteries were dug up and examined, there was nothing left of them; they were fully decomposed. Yet the

power continued to flow from them?

http://www.hbci.com/~wenonah/history/nathan-s.htm

Most recognized that electricity was simply a by-product or epiphenomenona of a more fundamental

agency, which entered the grounded lines. Rheostats somehow "tuned" the potentials of this earth energy.

While Reichenbach discovered the fundamental permeating nature of "Od force", several others showed the

essential unity of earth energy and the human aura.

It was found possible to "match and tune" these energies through the use of rheostats and capacitors.

Persons who were weak and infirm actually experienced vitalizing elevations when connected to the

ground energy through these special rheostatic tuners.

Tesla believed that ultra fine corpuscles from the sun permeated the entire earth, manifesting as static

charge.

13

Tesla further conjectured that these rays came primarily from the sun, since it was ejecting matter "at

excessively high voltages".

If this were so, reasoned Tesla, then sunlight contained something of this electro-active component ... and it

was certainly possible to derive electrical energy from sunlight.

Tesla also demonstrated the extraction of free electrical power from solar energy.

A grounded mica capacitor is surmounted by a highly polished zinc plate.

This plate may be poised in a highly evacuated glass container to best advantage, the zinc not exposed to

corrosive influences.

The tube is elevated and exposed to sunlight.

The mica capacitor is connected in series with the vacuum tube.

After only several minutes of exposure time, the stored electrical energy is formidable, producing a

powerful white arc discharge.

Tesla patented this device.

Mr. Stubblefield developed a peculiar bi-metallic induction coil which, when buried, draw up sufficient

electrical power to operate lamps and other appliances which he designed and tested.

A great length of both cotton-insulated copper and bare iron wires were wound together in a "bifilar"

arrangement on a large iron stove bolt.

The windings were held side by side throughout the coil.

His patent specification describes the device as a "terminal, which draws electricity out of the ground".

Stubblefield shared this particular fact with only one person.

I spoke with an academician who had the extreme privilege of speaking with Mr. Stubblefield's son,

Bernard Stubblefield.

Bernard, by this time himself quite aged, told that his father's method in locating the "right spot" was

deliberate and time consuming.

His father referred to the device as a "receptive terminal" and not a battery.

Despite the insistence of Patent Officers in calling the device a "battery", Stubblefield declared it to be an

"energy receiver ... a receptive cell for intercepting electrical ground waves".

Its conductive ability somehow absorbs and directs the enormous volumes of earth energy.

The induction coil, which bears his name is equipped with three coils which are wrapped around upon a

heavy iron core.

Bare iron wire and cotton covered copper wire are wrapped side by side, comprising a primary coil body.

Each layer of this primary coil body is covered by a band of cotton insulation, bringing four wire leads to

the coil terminus.

14

Two leads of iron and two of copper are external to the coil.

Commercial electrical power is obtained through these connective terminals.

Furthermore, though the Stubblefield power receiver is wound like an induction coil, it produces a steady

direct current output.

This poses additional problems for the conventional engineers.

Electrical induction only occurs with electrical alternations, oscillations, and impulses.

Witnesses described ground-powered motors which ran unceasingly and unattended for months without

need for replacing or replenishing the ground battery.

Small machinery, clocks, and loud gongs were run by other ground-buried cells as reported by credible

witnesses.

Stubblefield may have discovered the auto-magnifying voltage effect of electrostatic induction in coils

before Tesla, who later utilized the effect in his special electrostatic Transformers.

Nevertheless, different aspects of this ground sensitivity were discovered and differently implemented

throughout the following years.

T.H. Moray (1935) also discovered long-range articulate tuning through the ground from a fixed single site.

His "radiant energy listening device" permitted him to scan a tract of land and actually eavesdrop on distant

conversations and sounds through earphones.

This device did not implement a microphone.

The Moray Listening Device used a grounded rod and special large germanium detector.

How does a stationary tuner sweep across land and pinpoint sound sources?

Stanley Rogers (1932) discovered the same long-range scanning effect when, using a radionic tuner for

mineral detection, he found it possible to sweep a field or meadow with a variable capacitor.

Adjustments on these grounded tuners could sweep across land, revealing and mapping every mineral

contour.

Dr. R. Drown (1951) independently developed a compact device, which could sweep, scan, and delve

through subterranean grounds for the specific purpose of ore detection.

This device permitted photographic detection of ores swept through the ground, isolating specifically

sought mineral deposits.

NOTE: Sources indicate that this type of long-range voice monitoring technology is now employed and is

in use. Areas can be tuned in and monitored using the technology first demonstrated by Moray and

Stubblelfield.

SUNSET

I was the quite fortunate recipient of an unexpected personal letter while writing my original treatise on

Nathan Stubblefield.

15

It was told by a gentleman who received the account through a man who witnessed the following.

Neighbors had not seen Nathan for several days.

As they were worried about his health, they attempted to call on him.

The lock was secured from the inside.

It was a lonely, cold, and rainy March day when old friends and neighbors broke the lock on Nathan's cabin

and entered.

He had passed away in his bed, the probable victim of malnutrition and fatigue.

They all noticed that the interior of the cabin was "toasty warm", as if heated by a strong fire.

Moved to locate the source of this heat, town officials found "two highly polished metal mirrors which

faced each other, radiating a very great heat in rippling waves".

Now this, I must say, is a truly great discovery and last mystery.

It fulfills what Nathan reported in his last testimony.

Nathan's deepest confidence was in those kind and compassionate people who continued to seek him out

with love and concern to his last days.

Abandoned by all, he wished one of his dearest neighbors to write a biography.

Perhaps he wished to explain his life, an apology for all his ways.

He said, "I have lived fifty years ahead of everybody else".

While often sounding inspirational, these are words of deepest sorrow.

To live with a vision of the future is to experience the surprising, often disappointing rejection and

resistance of all who surround.

Some said he was incapable of loving others.

But ... it was love, his love, which coaxed the living sunshine out of hard, rocky ground ... the resounding

waves of an eternal subterranean sea of energy.

Final years

Stubblefield later lived in a self-imposed isolation in a crude shelter near Almo, Kentucky and eventually

starved to death on March 28, 1928.

He was buried in the Bowman Cemetery in Murray, Kentucky.

16

NOTE: Could we consider that Nathan Stubblefield was so far ahead of his time that we have yet to

properly figure out all of what he discovered? Was he only 50 years ahead of everyone else? Or is it more

likely that he was about a century plus ahead everyone else?

Consider this: Stubblefield did not know where the power for his experiments was coming from, and yet

he was able to heat his house to toasty warm levels using free energy.

Michael Faraday and Lester Hendershot look to be the only people who were looking at planetary motion

as the source of free energy.

Today we understand that it is the high-speed motion of our planet that provides the vast limitless amounts

of free energy.

That's what this write is all about. To share the Joy of our awakening to the truth that has been kept from us

for such a long time. This So that we can inspire fellowship to bring on line the technology of life that the

early explorers into free energy first mapped out for us so long ago.

Dr. Mallove's thoughts in 2004: "Unlimited clean free energy is only a few months away."

Nathan Stubblefield proved it, Tesla showed it to the world; It's our turn, how about we build it?

· Chapter three

· Nichola Tesla:

"Ere many generations pass, our machinery will be driven by a power obtainable at any point of the

universe.

This idea is not novel. Men have been led to it long ago by instinct or reason; it has been expressed in many

ways, and in many places, in the history of old and new.

We find it in the delightful myth of Antheus, who derives power from the earth; we find it among the subtle

speculations of one of your splendid mathematicians and in many hints and statements of thinkers of the

present time.

Throughout space there is energy.

Is this energy static or kinetic! If static our hopes are in vain; if kinetic — and this we know it is, for certain

— then it is a mere question of time when men will succeed in attaching their machinery to the very

wheelwork of nature."

As we see, at that early time it was conventional engineering theory to accept that energy can only be

extracted if a field is in motion.

That theory of operation, only kinetic fields can be accessed for energy extraction, is still somewhat

regarded as conventional notion of extracting energy.

What is understood now and may have been understood by Dr. Mallove some time early in 2004 or

possibly in 2003, is that the static field can be accessed for energy. Not only can the static field be accessed

for energy, it contains within itself the largest possible source of free energy. Can we wonder if Henry

Andrew Bumstead by 1920 had figured this out?

17

There are four elements to extracting energy from a static field. This brings us into the area of relativistic

science. Dimensions, split fields, high-speed motion, and time differentials. The four elements of free

energy.

All previous machines that do work for us are based on differentials of one kind or another.

Differences in temperature, voltage or pressure, can be used to power a device.

It is the differential between the two sides from the energy input source to the output that allows our

machine to extract useful energy do its work.

Previous machines use kinetic actions to access the differentials that provide their motive source of useful

work.

To extract energy from static fields, what type of differential will we need to use to extract energy from

them?

Shall we take a look at the theoretical explanation to help us understand the how and why that has been so

difficult and confusing to bring understanding for so long?

Here then are the four elements of free energy.

The four key points to understanding free energy..

Here's a post from 2010 that explains the basis of Velocity power sources.

· Chapter four

· The Four Elements of Free energy

Speed Torque Fields Time

Patrick Sullivan

The Four Elements of Free Energy

Revised August 24, 2010

Nicolaus Copernicus (19 February 1473 – 24 May 1543) was a Renaissance astronomer and the first person

to formulate a comprehensive heliocentric cosmology, which displaced the Earth from the center of the

universe.[2wiki]

Copernicus' epochal book, De revolutionibus orbium coelestium (On the Revolutions of the Celestial

Spheres), published just before his death in 1543, is often regarded as the starting point of modern

astronomy and the defining epiphany that began the scientific revolution.

His heliocentric model, with the Sun at the center of the universe, demonstrated that the observed motions

of celestial objects can be explained without putting Earth at rest in the center of the universe.

His work stimulated further scientific investigations, becoming a landmark in the history of science that is

often referred to as the Copernican Revolution.

18

The question of planetary motion was several thousand years in the making until the proof of it finally

arrived in 1729.

The ancient Egyptians were of the flat earth model of our planet. Planet earth stood still while everything in

the Cosmos revolved around it. Case closed for their knowledge in the area of planetary motion.

IN the 5th century BC the Hellenes, the people of Greece, were the first reported to consider the question of whether or not our planet is in motion.

Others speculated at that time that indeed, we are in motion. No proof, as how could it be proved one way

or the other?

Greek astronomer Claudius Ptolemy (90-168AD) deduced that our planet was not in motion as if it was,

then the wind would be blowing the trees about constantly. Seems logical enough, doesn't it?

The question of earth motion awakened in 1543 with the Polish astronomer Copernicus making a

mathematical deduction that planet earth was in motion.

Astronomer Giordano Bruno (1548-1600) claimed that planet earth was in motion, and paid the ultimate

price for his mistake of defending his position. "Farewell Bruno." Did they in some strange way love you?

Galileo Galilei (1564-1642) was a proponent of heliocentrism; our earth is in motion around our sun. He

was shown the instruments of torture by the inquisition, and thereby brought to a change of mind.

He came around to their point of view and Agreed that the earth stood still and was not in motion.

Is it a curious question to consider why the authorities would stand for so long with the position that the

earth was not in motion? What could be so potentially dangerous to the authorities that they went to such

great lengths to make the case that planet earth stood still?

Here's a thought that jumped into this head about an hour ago concerning this issue.

If we were on a ship traveling the wide-open seas would we as individuals relate to each other differently than when we think of others who are spread to far distant places about the planet? That is, would there be

a difference in how we may view our fellow human beings if we are all on the same ship traveling the

wide-open seas?

If we think of earth as carrying us all through space and time together would we see our fellow passengers

as in the same boat as us?

Here's something that may give us a view about this issue. This writer has written many articles concerning

the extraterrestrial intervention into our world. From researching the field I have surmised that our

dominant classes have known of the presence of the high level powers of ET for many centuries now.

The Being known to us as the Infant Jesus of Prague that made his appearance in 1648, may be one of our

direct relatives.

His figure is found in sculptures, carvings and paintings from the year 1350 AD. That was when his

gracious presence was first recorded by artisans of the time.

He contacted the centers of power, the Royals, Vatican and financiers. They declined the opportunity to

interact with Maxmillian the representative sent to us from the Galactic Federation of light.

19

Upon learning of this secret history of contacts between the high level extraterrestrial powers and our

dominant classes here on planet earth I was stimulated by curiosity and felt duty bound to ask the question:

"Why did our earth powers reject contact with an advanced species of high level powers in 1350 AD?"

The answer from Maxmillian was simple and direct: "They had already perfected the predator relation

ship."

Joan of arc was burned at the stake in 1431AD. About 80 years after the 'centers of power' on planet earth

knew of advanced extraterrestrial societies. Despite knowing of the high level extraterrestrial powers and

their non-use of force, the centers of power here on planet earth continued to burn people at the stake?

Could the powers that be when confronted with the issue of planetary motion have thought through what

may be the response in the ordinary people? Would the notion of riding on a ship traveling through the

Cosmos provoke the ordinary people to consider the consequences of what could happen with letting others

decide their fate?

Why are so many dieing unnecessarily of disease and living in poverty while many others seem to seldom

notice? If we were all in the same boat, would we let this be? Does the fact of planetary motion give us a

common ground to stand on? If the ship goes down, aren't we all going with it? Why did they build so

many underground shelters?

On the good ship lollipop it's a nice trip to the candy shop where bob-bons play on the sunny beach of

peppermint bay.

Is there a sense in this planetary motion that could give us some clues of how far away our purported leaders are from honesty with us? What's this enterprise about anyways?

Consider this: Its six hundred and sixty two years since the advanced powers introduced themselves to our

earth leaders, how have we let the rascals get away with their capers for so long?

As to the question of planetary motion, it was eventually accepted by the community in 1729 when Sir

James Bradley (1693-1762) proved by the aberration of light that it was true, our planet is in motion all of

the time.

Have you ever heard of the aberration of light?

Are you familiar with the Doppler shift as it is related to sound waves?

For instance, a train is approaching blowing its whistle loudly. As it goes by on past where you are standing

and listening, the pitch of the whistle changes.

The aberration of light is somewhat analogous to this effect. It bends depending on which way you are

going in relation to it.

Bradley marked out several stars and monitored their position. At different times of the year the stars varied

a slight degree in position. Think of light as you're heading toward the star. At a certain point as we rotate

around the sun we then head in the opposite direction, away from the star. The star will show a slightly

different position during that time of travel.

Bradley deduced that the explanation for why the star changed position was due to the direction in which

our planet was moving at the time. That was His proof of planetary motion that was finally accepted, nearly two centuries after Copernicus proved it mathematically.

20

As with sound waves where we noted a change in pitch depending on which way the sound was being

delivered to us, with light waves there is a change in position as we make observations of the stars.

Michael Faraday in 1832 knew the rotational speed of planet earth. With as much information as we have

to day, how many of us are aware of our high-speed planetary motion?

How could this simple fact that we are traveling through space at high speed, have missed most of the early

researchers in considering where the free energy is coming from?

Here is something that is another one of those curiosities; the ancient Egyptians had noted stars making

small circles in the sky and considered that the starts were moving about.

So could we consider that they had already discovered the aberration of light, yet had no question to ask as

what it indicated that they were observing?

What James Bradley would demonstrate in 1729 to be the proof of planetary motion (circular motion of

distant stars) was known to them, yet they never considered whether or not our planet was in motion to

begin with?

The above is presented for your consideration. If you formulate a good thesis about the questions involved

there, will love to hear them.

The observation was made of distant star motion by the Egyptians, though no theory was put forth to explain what it was?

Did No one question at that early time that the earth could be in motion?

Why was this? Was a motionless earth a settled question that blocked further inquiry into the issue?

Could this sort of thinking help explain why the static fields of our planet have not been considered as a

place to extract energy from? Classical engineering only allows for kinetic fields to be used for power

sources. Could there be similarities in our response to Accepted dogmas that have proved to be hindering

our understanding of things? Is it even odder that it took so long (about 2500 years) since the question was

first asked if we are in motion or not, to prove that our planet is in motion?

That our solar and galactic systems are also in motion along with planetary rotation around our sun was not

known until the 20th century when Edwin Hubbell (1889-1953) drew those first startling conclusions from

his astronomical observations.

So earth is orbiting our sun. Our solar and galactic system is also in orbit with us.

This combined high-speed motion giving our earth nearly 600 (precisely 559.4 according to some) miles

per second travel through the Cosmos is the key to powering our free energy machines.

This is the First Element of Free Energy; planetary motion.

Here's a little review of where we are at now:

The planet we call home (Earth) is sailing through space at close to 600 miles per second.

Our entire planet including the atmosphere is speeding along with us and so there is little clue that we are in

motion at all.

Copernicus stated that our planet is in motion, though proof of his observations required another 186 years

of astronomical study to find acceptance.

21

Copernicus wrote in 1543: "The apparent retrograde and direct motion of the planets arises not from their

motion but from the earth's.

The motion of the earth alone, therefore, suffices to explain so many apparent inequalities in the heavens."

James Bradley was the first to have his proofs accepted that we are in fact in motion.

The discovery of the aberration of light was, for all realistic purposes, conclusive evidence for the

movement of the Earth, and hence for the correctness of Aristarchus' and Kepler's theories; it was

announced to the Royal Society in January 1729

Free energy, High speed Planetary motion; That's the basis of it all: Power of the Cosmos. It will provide

us forever with all of our energy needs. Does the proof of the availability of free energy Absolutely end

forever the notion that we need to engage in resource base competition with others?

While on this question of earth motion and how peering into the night skies with telescopes and lenses it

was proven that we are in motion; the element Helium was first discovered by observing a solar eclipse in

1868. Helium was only discovered on earth years later in 1895.

Is there something here that clues us to looking far away to discover things right near to us?

The next technical issue to consider is how do we attach our machines to the wheelworks of the Cosmos?

Dimensionality is the Second element after earth motion to consider in our 4 Elements of free energy.

"Where's the torque"

How do we develop and produce torque in our Velocity power source machines?"

"Where's the torque point?"

In a windmill the main torque point would be where the machine is attached to the ground. In a automobile

the main torque point for our engine to deliver rotary motion to our transmission would be the engine

mounts, that attach the engine to the frame of our car.

To aid us in understanding the Second element required for extracting free energy, we can first take a look

at the question of dimensions and how they relate to it.

Consider this: We Move our bodies (Up) then (Down)----(Left) then (Right)----and finally (Forward) then

(Backward.)

We have now traveled through 3 dimensional spaces. These are our three ordinary dimensions.

Length, width and height represent mathematically the first three dimensions.

Now if we decide to dance around in a pickup truck bed motoring up the street we will feel the wind in our

faces and still have our three dimensional motion; the truck's motion itself is now our fourth dimension.

August Mobius (1790-1868) was the first to mathematically define the fourth dimension working it through

between the years 1824-1827.

So as we dance around in our 3 dimensional world here on planet Earth, we are at the same time flying

through space on the forth dimension, Earth's high-speed (Remember 600 miles per second) orbital motion.

An immediately noticeable difference between riding in the pickup bed of a truck and our planet's motion is

that there is no 600 miles per second wind blowing into our faces. The reason we do not feel anything is

because the atmosphere is in the same relative motion as we and everything else is on our planet.

The term often used is called "Frame dragging." The first three plus the forth dimension then is our original

inertial frame of reference, with everything within it moving along at the same speed, the same relative

motion.

So we could consider that we are ordinarily operating in a four-dimensional world, able to detect easily just

three?

In this instance then with our three dimensional motion moving about in the bed of a pickup truck, the truck

is giving us our fourth dimensional motion, The wind in our face as we travel in our open truck then could

be looked at as our fifth dimension.

Theodore Kaluza (1885-1954) worked for years to produce the mathematics of the 5th dimension that he

gifted to us in 1921. This 5th dimensional aspect is the first half of the Kaluza-Klain theory that enabled us

to identify and see the torque point that will allow us to tap into the power of the Cosmos.

If we install a windmill in the bed of our pickup truck and drive up the street the air flow passing the blades will turn that frictional airflow into motion that could drive a generator for energy.

When we attach our windmill to the bed of the pickup truck would it be somewhat, if ever so slightly,

analogous to our extracting energy between the 4th and 5th dimension? The pickup truck in this instance

as our 4th dimension and the wind as our 5th dimension in this case? See the connection with the 4th

dimension of Mobius in 1827 and the 5th dimensional mathematics of Kaluza in 1921?

Driving up the street our pickup truck gave us our 4th dimensional motion, and the wind is now our 5th

dimension.

Oh, there are a couple of other technical issues besides dimensionality to think about when we change from

a windmill in our pickup truck to the motion of planet earth itself. First, is there wind in space?

The Michelson–Morley experiment was performed in 1887 seeking to detect the theorized aether of space

that we are traveling through. It returned a null result, and did not find any aether. No interacting field

strength was detected. This was the first experiment that eventually leads to the beginnings of the theory of

relativity. Time is not a constant; rather it is variable with the speed of its motion.

As there is no wind in space, there is also no friction point available to torque at as a windmill operating in

our atmosphere; so how are we supposed to spin our planetary free energy windmill to extract our free

energy?

This brings us to our 3rd Element of free energy.

The dual field nature of the thing we call electricity.

Electricity, and how it can be used to extract energy derived from the high-speed motion of the planet....

lets not forget, we are constantly sailing through space at 600 miles per second. Could that indicate that

there is a whole lot of (24/7) free energy just waiting out there for us to tap into?

Have you checked your electric bill lately? You may notice that you are billed for the amount of kilowatts

that you have used since the previous meter reading was sent to you. Multiply Amps time's volts gives us

23

the number of our watts. Multiply again by One thousand watts and we have our kilowatt. One kilowatt that

can cost usually between .8 cents to about .20 cents a kilowatt hour.

Electricity is considered to travel at the speed of light. (186,000 miles per second), Faraday measured it at

144,000 miles per second passing through a wire.

When a vacuum tube or a semiconductor passes electricity, the 2 sides (amps & volts) of the electric field

are split due to the heater in the tube causing the electrons to separate and pass to the plate separating the

magnetic from the electric side. Thermionic emission is the term used. It's what Michael Faraday did in his

lab when he heated a wire with electricity it. The volts disappeared, while the amperage remained. This will

create a diode effect, turning an alternating current signal into a direct current pulse. The electrical field

will flow in one direction.

A semiconductor produces a similar effect, using its unique properties. The volts side will continue on at near speed of light while the amperage (phonon) side will travel through a semiconductor at the speed of

sound.

Another way to describe Electricity (both sides of the combined electric magnetic fields) is to consider that

these fields (magnetic and electric combined together to make this thing we call electricity) have specific

and particular types of particle spin.

The amperage (magnetic) side is called full spin particles (360 degrees of motion) and these produce a

gyroscopic effect. The full spin side is the magnetic force carrier side of our electricity. Part of the reason

for describing the two sides of electricity with spin is because it is something that is familiar to us and may

help to clarify what it is that we need to do to be able to design and bring on line free energy machines of

life.

Volts (static electricity is usually associated with this side of electricity) are the measurement tools to help

us define our half spin (720 degrees of motion) electron particles.

While the speculation to why they are half spin concerned the interaction with the standing waves of our

planet, it may be more in the area of what the universe oscillates at that is producing this affect.

There is a top and bottom spin associated with the half spin particle; 360 degrees goes from top to bottom.

The particle needs another 360 degrees to return to its original position. That may explain why there needs

to be 720 degrees of rotation for the particle.

It enters into the area of beyond the need to know as to how to build and operate velocity power sources, a

distance beyond into deeper theoretical areas.

Volts are our electron side (negative). Amps are our phonon, magnetic side (B+ the positive).

While all particles are considered to be of zero mass, in a static field reaction a volt could be considered to

have a mass of 1 and on the other side of the circuit a B+ magnetic particle could be considered to have

3600 mass units. The added dimension (5th) reaction is where the mass difference may be measured at.

The mass numbers are theorized from the size of the (Proton 1800 emu), (Neutron 1800 emu) and (Electron

1 emu). An atom of mass is considered to be about 3600 emu.

All matter is electrical in nature. All matter for our purposes can be considered to be of 3600 electron mass

units.

24

The understanding of extracting Free energy from high speed planetary motion and how it is accomplished

concerns how the different sides of the electric field interacts with the standing waves of our planet.

These standing waves are also called "Zero point fields" "The static field" or "planetary inertial fields."

Bose-Einstein (Bosons-full) and Fermi-Dirac (Fermions-half) statistics explain the mathematical basis that

comprises the statistical variations between the 2 sides of the electric field.

Electricity is comprised of these two prime components; light and heavy particles.

Both sides of the electric field, half and full spin particles, also exhibit wave particle duality.

Electricity is a section of a part of the electro-magnetic spectrum.

The magnetic side is the amperage full spin particle side. These are the force carriers in our machine. This

is the side that travels at the speed of sound when split by a vacuum tube, semi conductor or a high voltage

transformer.

While Michael Faraday determined electricity flowed at 144,000miles per second in a wire, could we

consider that he was measuring an average speed for all particles in the flow of electricity?

The mystery of Ferro-resonance and how understanding it helps to explain the speed of light and sound;

dimensions, high speed planetary motion and most everything about electricity that has been a mystery for

so long.

What is Ferro-resonance anyways?

It is Something discovered by power engineers many years ago when they were switching power lines, if

there was an imbalance in the voltages between them, transformers, some of them far away in the lines

somewhere, could explode in the most violent of fashions, causing great alarm, distress and dollars and

cents damages.

This can also happen when a relatively small amount of energy is used from a power line. For instance, a

100 amp welding machine operating for several hours continuously to thaw water pipes, has been known to

cause a large transformer to overheat and explode shutting down all of the other users of that line.

Power engineers are well aware of this and have an organization that researches the issue to prevent it these

situations from happening. It's well understood and proper procedural precautions are in place to prevent

these sorts of things from happening now.

There is also something called hysteresis that can happen in an electric motor. Industrial units have

precautions noted in their operational manuals to not store bundles of wires nearby as it has been known to

produce overheating of motors by inductive effects producing hysteresis.

There is another potential overheat to electric motors, called eddy currents. In the early days of electricity it

was believed that eddy currents were in the air and somehow entered into motors and overheated them.

Motors are now designed to reduce this effect.

What the theory of operation of Velocity power sources in this book is about, makes understandable that

the events associated will all of these; ferro-resonance, hysteresis, and eddy currents, are all static field

reactions due to separations between the two sides of electricity.

In a high voltage power line, when the line is disconnected, the high voltage static side, will collapse at the

speed of light. The amperage side will flow at the speed of sound. This difference between them opens up

the path for the amperage to be disconnected from the high voltage, that snapped at the near speed of light

25

and temporarily departed the original frame of reference (4th), and allowed the amperage to torque at the plane of the dimension and enter the 5th dimension.

That is the explanation for why so much energy pours in. It's coming from the motion of our planet.

Hysteresis, eddy currents as with ferro-resonance; the separation between the two sides of the electric field

allow for free energy to pour in.

Volts are the electric side that is comprised of half spin particles.

Have you ever held a bicycle wheel in your hands and had someone spin it? If so, did you notice how it

tends to want to stay in its original position?

This is the gyroscopic effect. It's an inertial reaction that is felt when attempting to change its original

position.

In particle spin, the full spin particles operate like a gyroscope. When a coil is energized these full spin

particles attach to a motor or heater coil where they then gyroscopically latch that coil to the standing

waves of our planet, the static field.

It is the magnetic side (full spin) that remains in our heater or motor coil attached together; coil and

standing waves.

The volts and amps, half spin and full spin, are first installed in our heater coil. The amps remain

gyroscopically coupling our motor or heater coil to the standing waves of our planet.

The half spin, volts side of our circuit, upon acceleration into percents of the speed of light, produces a

mass absorption effect that changes the rate of time flow (slows it) that then produces a change in dimensional position. A shift in time is what does it. The half spin act as if they are time compressed. Time

flows at a slowed rate.

A relativistic time shift.

The 4th element of free energy is about how the effects of acceleration relate to the Theory of relativity.

Time compression in one side of the combined fields is what gives us our differentials in our machine that

allow us to power up our free energy devices.

The Michelson–Morley experiment was performed in 1887 by Albert Michelson and Edward Morley and is

called the Most famous "failed" experiment

What was theorized early on and has become known as the Einstein-Fitzgerald-Lorentz Contraction

concerns how electric fields contract when undergoing acceleration.

The contractions in electric fields slow the rate of time flow when undergoing acceleration.

While it was originally called the theory of relativity, today time contraction depending on field intensity

and strength is a generally accepted fact of science.

Most of Our current industrial Machines operate based upon principles of differences in temperatures,

pressures, and voltages.

To extract the free energy of high-speed planetary motion (The Power of the Cosmos) we require machines

that operate on differences of time.

26

Here is a description of how we can draw energy from the planet's motion: First by separating the 2 sides of the electric field (magnetic from electric) and inserting one side into our heater coils,(full spin gyroscopic

particles, magnetic) we then accelerate the other side (half spin, volts) while maintaining the magnetic in

the heater coils.

For instance with our half spin particles, A coordinate acceleration that reaches 80% the speed of light will

produce a corresponding clock rate time flow of 60% the rate of the original time flow.

So one side of our free machine (full spin) is showing us one time and the other side of the machine (half

spin) has an ever so slightly different time.

As the Earth is traveling at nearly 600 miles per second, The "Static fields" of our planet that are powered

by the high speed motion of our planet will pour energy into to our full spin particles that are installed in

the heater coils and gyroscopically attached to the static field, so as to maintain the original position within

their inertial frame of reference; 4th dimension, when the half spin side is accelerated and are time

compressed..

Because of the slight time differences between the 2 sides (electric field-magnetic field) the Planet's

inertial fields are opened up to maintain positional stability due to the standing waves produced by the

gyroscopic effects of the full spin particles that firmly engage the magnetic electric and static fields and the

heater coil. A full spin particle produces a magnetic standing wave (linked with the motor heater coil) and

our planet also has standing waves.

The magnetic full spin standing wave is rotatory and will roll and produce torque once the electric field is

removed by time compression. Because it is linked in the heater coil and the heater coil is firmly attached

to he forth dimension, with the standing waves-zero point-inertial static field, it will torque at the plane of

the dimension opening up the flow of planetary velocity power and energy into the heater coil.

It is the high velocity of our planet through space that allows for this. Velocity power sources; That's where

the name arises from. Planetary inertial field generators were the first name for them. The inertial fields are

what drive the free energy for our use.

Here we can review how to extract energy from the Power of the Cosmos.

These are the four basic elements of Free energy.

1).(Motion of planet) Nearly 600 miles per second.

This high-speed motion explains where the energy is coming from. No violation of the first and second

laws of thermodynamics.

The energy input arises from high-speed planetary motion as we sail through the Cosmos.

2). (Torque point: Extra-dimensional aspects to consider) A generator hooked to the bumper of our truck

giving us energy between the bumper and the roadway or a windmill mounted in the bed of the pickup

truck, or according to Kaluza-Klein, between the 4th and 5th dimension.

Torque is produced at "The plane of the dimension" between the motion of the planet (4th) and the

universe (5th dimension).

3). (2 Prime components of the electrical field) Our Light (half) and heavy (Full spin) particles that are the

two prime components of the electrical field.

27

The full spin particles produce a gyroscopic effect that creates a standing wave that then interleaves at the

sub atomic level with both our Free energy machine, latching heater motor coils to the standing waves of

our planet.

Tesla discovered three of these "Planetary inertial fields" in the early 1900s.

One researcher is reported to have identified 9 of these planetary inertial standing wave fields.

12 Fields are theorized to be associated with our planet.

These "Standing waves" are also zero point fields and planetary inertial fields; all components of the same

thing, the Static fields of our planet.

4). (Relativistic time contraction of half spin particles) light side of the electrical field.

When our half spin particles are accelerated, (resonant frequencies will do the job) and the clock rate of

time is slowed, Then as the Earth moves with our full spin particles in our Free energy machine, the full

spin particles attempt to roll, or "Torque" while in the machine coils.

The separation of the electrical field (Relativistic time shift between half and full spin particles) is what

causes this.

The "Static Field" (is the Composite of total standing waves of the inertial fields of the planet) pours energy in to maintain the full spin gyroscopically attached position within the original inertial frame of

reference, the 4th dimension.

The half spin particles are time compressed and have momentarily departed the original inertial frame of

reference and now exist in the 5th dimension.

Our free energy machine has produced a "Displacement in the time space continuum." This is how a static

field can be accessed for energy. Time is the differential in our machine.

Production of energy in this manner will produce Cold electricity. The electrochemical cells of an HHO

generator show a cold electrical reaction. They are also over unity free energy devices.

Cold electricity is proof of an over unity reaction arising from the interaction with the Static fields of the

planet.

These again are the 4 basic elements as to where to find and extract Free energy:

1). Planet motion:

2). Torque point:

3). 2 prime electrical components:

4). Acceleration producing relativistic time shift:

Free energy: "The Power of the Cosmos."

What are we waiting for? Lets build some real big machines and quick.

http://en.wikipedia.org/wiki/Nicolaus_Copernicus

http://en.wikipedia.org/wiki/James_Bradley

http://peswiki.com/index.php/PowerPedia:Paul_Pantone

· Chapter five

· Cold Electricity

"What it?"

Nichola Tesla (1856-1943) the inventor and creator of the first alternating current system of electrical

distribution and also the inventor of the poly-phase motor that set the standard for electrical generation and

distribution that is still in use today also was the first person to describe a strange and mysterious property

of electricity called "Cold electricity."

In1899 while in his laboratory experimenting and building a wireless light bulb, Tesla quite accidentally

discovered that the apparatus that he had built, an early example of an 'over unity device' was causing the

light bulb to glow, but itself remained cool to the touch.

This was something quite new and unexpected. Normally all electrical fixtures heat up from the electrical

flow that passes through them when in operation yet in this instance, the opposite happened; the fixture in

which the bulb was glowing did not get hot; instead it became cool to the touch.

Tesla noted this puzzling, strange and unexpected development while observing the light bulb glow hot

while the apparatus remained cool and he named this affect 'Cold Electricity.' Tesla was the first though

not the last to discover and record some of the curiosities and complex properties surrounding his discovery

of 'Cold electricity.

'Tesla was asked by a reporter "Where does the energy that is lighting your light bulb and powering your apparatus come from?" Tesla replied, "I don't know, but I'm glad its there.

"Tesla continued to work on his over unity apparatus and had drawn up a proposal to build a device that

was capable of lighting 5000 light bulbs all at the same time without any external connections to it.

When his patron and Baron of the finance world J.P. Morgan received word of what Tesla was working

on, he immediately ended his funding and had his workmen tear down the workshop that Tesla was doing

his research in. So ended the revolutionary early work of the electrical genius Tesla and with it, the end of

the 'free energy' movement of its day.

"Cold electricity' remained only a mere laboratory curiosity that was forgotten and not to be heard from

again until its rediscovery in the laboratory of the next genius of Peace and prosperity, the electrical wizard

of Salt Lake city, Utah, Dr. Henry Moray.

Thomas Henry Moray (1892-1974) as a young man spent a great deal of time upon his hobby of building

crystal radio receivers. From his work with crystals he was able to build a radio receiver that had an

amplifier so strong that it could shake a speaker at high volumes.

With this intuitive knowledge he began to build amplifiers that could draw electricity from the "Sea of

Energy," as was the name that he gave to one of his books.

Using his crystal generator he was able to create useful power to operate lights, heaters and with

modifications, motors.

One of the 'crystal controlled' amplifiers Dr. Moray built could fit on a kitchen table and would produce

50,000 watts of electricity. In dollars and cents, depending on where you live, that is about $120 a day

29

worth of electricity, not bad considering that Dr. Moray was like Tesla, not really sure where this energy

was coming from. A reporter posed the same question to Dr. Moray that was asked of Tesla: "Where is the

energy coming from that is generated by your machine?" To which Dr. Moray responded: "I don't know,

but I think it has something to do with the different sizes of particles."

Dr. Moray's machine, as was Tesla's, an over unity device, that is it collected energy from the Cosmos and

turned it into useful electrical and heat energy. His 'crystal' generator was creating energy from a vast

source; 'The Cosmos.'

Here Dr. Moray comments on an aspect that he noted during a test of one of his 'free energy' machines that

while his generator was operating, the method of energy production produced a different type of electricity.

http://en.wikipedia.org/wiki/Thomas_Henry_Moray

Dr. Moray's words: "That the size of wire in the transformer could not carry the amperage passing through

it without burning up, if ordinary current were used, yet the wires remained absolutely cool, no matter how

long the machine operated."

Here was Dr. Moray making the same discovery that Tesla had made nearly 30 years before; the fact of the presence of "Cold electricity" in his over unity machine.

Of the many over unity machines that have been built since then, cold electricity is a fact of life in all of

them and has been noted repeatedly by experimenters such as the late Edwin Gray (1925-1989).

"So, what is Cold Electricity? Why is it only found in over unity machines? Will Cold electricity produce

light, heat and run motors the same as ordinary electricity does? Can I put a meter on it and sell it?"

Let's start with the last question first: "Can I put a meter on it and sell it?"

The answer is yes. The energy (Cold electricity) itself is a gift of the Cosmos and is free, but to access it

you will need a specialized type of machine that can extract Cold electricity and that will cost money and

yes you could put a meter on it and sell it.

The good news though, if you don't own your own machine and instead receive it from a metered source,

you will be able to get a real large amount of energy for a very small price in comparison to what you pay

for energy now.

Cold electricity can also be delivered and transmitted through the existing electrical grid system.

Next to last question: "Will Cold electricity give us heat, light and power our motors?"

"Yes, Yes, and Yes. Tesla, Dr. Moray, and Nathan Stubblefield (1860-1928) and many others have all

produced light, heat and motor operation using cold electricity.

Stubblefield heated his home using free energy though he may have not been aware of the nature of Cold electricity. Tesla was the first to have given a name to the strange phenomenon of Cold electricity.

http://www.icehouse.net/john34/stubblefield.html

Now that we know that Cold electricity will do everything that we need done, it will Light and Heat our

homes and industry, and run as many motors as we want to run, all using free energy; cold electricity is

truly The Power of the Cosmos.

30

To explain how it is created we will need to examine a few facts concerning the science of Relativity and

its relationship to hyper dimensional physics.

Stay with us, we need you. This is a bit far reaching at first, yet understandable once some thought is given

to it.

First let's talk a little about ordinary machines that produce energy.

Take a windmill for example, the wind is in motion and the windmill is standing still. A 'frictional' effect

between the blades of the windmill and the wind produce a region of high pressure on the front edge of the

blade and a low pressure on the backside.

The windmill is fastened to the ground; this connection to the ground is the primary 'torque' point between

the wind and the windmill generator that allows energy to be extracted.

As the wind blows, High pressure-air pours onto the blades venting to the low-pressure side producing

rotary motion.

The primary 'torque' point, our ground connection where the windmill is attached, allows the wind energy

to produce and transfer the rotary motion to our generator.

With the wind, we can feel it upon our faces when it blows because we are at a different speed than the

wind.

The difference in speed between the wind and the surface of the Earth is the method that allows for energy

production.

We could say between a 'High' and 'Low' pressure zone, is where the energy can be extracted. In a

windmill, pressure is the differential. Wind is a kinetic field. .

A conventional automobile engine works on the same basic principle; High pressure is developed in the

combustion chamber by the expansion of gas from the burning of fuel and vents through the exhaust pipe to

the low-pressure zone, the atmosphere.

The pressure differential is what allows energy to be extracted. The crankshaft turns pressure differences

into rotary motion.

The main torque point is found in the engine mounts that attach the engine to the frame that allows the

rotary motion to drive the wheels.

The Earth is in high-speed motion (600 miles per second) through the Cosmos, though we cannot sense this

by ordinary methods because we are in motion at the same speed.

The technical question that is at the root of it all and will allow us to understand Cold electricity is "Where is the torque developed to power our machines from the high speed motion of the planet?

With our windmill it is easy to see because we securely fasten the windmill to the Earth. We can feel the

wind blow.

Our primary torque point, connection to earth, is seen and is obvious.

Unlike the windmill, that is fastened to the ground and allows us to take energy from the speed difference

of the wind and the motionless, in 'relation' to the wind, the solid earth, everything on planet earth is

moving at the same speed as our free energy machine, so where do we connect our machine to allow us to

31

tap into the high speed motion of the earth and find our 'Cold Electricity?' This is the dimensionality

aspect. The 4th to 5th dimension is our extraction point.

Applying the water is similar to electricity analogy consider An alternator as is on our cars and trucks is a

'pump,' of 'ordinary electricity' that supplies useful power. Most vehicles manufactured use 12 Volts and

(For ease of calculation) provide 100 Amperes, or Amps, of output. 12 Volts times 100 Amps = 1200 Watts

of power, or 1.2 Kilowatts of power, that is about .12, or 12 cents worth of electricity per hour.

'Cold Electricity' on the other hand does not come to us from an electrical generator, instead it comes from

the energy fields that surround us; "The sea of energy" as Dr. Moray called it.

The presence of 'Cold Electricity' is an indication that the machine is operating as an over unity device and

that the energy generated is being extracted from the 'Vacuum of space.'

Rather than ordinary electricity that is 'pumped and pushed' into a machine, cold electricity is drawn, like

in a vacuum, from the energy fields that occupy the entire universe. The connecting point is with our static

fields.

Dr. Moray had suggested as to where the energy is coming from in his over unity machine, we recall:

" I think it has something to do with the size of the particles."

Cold electricity in brief, is proof that the energy found in our machine is coming from the Cosmos.

The reason it is cold, is because the energy is coming into our machine from the universe itself. The wires

leading to a heater are not hot, though the heater coil is hot. This is because the energy is entering right in

the coil, rather than being pumped into the machine from the power cord.

· Chapter six

· Gabriel Kron

" My work is so beautiful that I wouldn't give it away for a million dollars."

Negative resistance

Multi Dimensions

The inertial frame of reference and the multi dimensional aspect of physics was examined by Gabriel Kron

who accidentally created an over unity "free energy" circuit.

One of Kron's electronic circuit designs was installed in 1963 into the Minuteman missile and was only

discovered to be over unity when the electronic control circuitry of the missile burned up while on station.

The Kron circuit was found to produce 7% over unity, that is 100 watts of energy in and 107 watts of

energy out.

After this the Kron circuit was clamped down with grounding circuitry to prevent the over heating due to

the over unity of energy that the circuit demonstrated.

Kron was able in the laboratory to tweak the circuit to 15% over unity.

He stated that much if his work was intuitive, and he did not know at this time where the energy was

coming from, only that it was there.

32

General Electric had the Kron circuit made secret and so it remains until this day; secret.

He developed what Dr. Henry Moray had also discovered and built, a negative resistance circuit.

In 1934 he joined General Electric and worked there in various departments, all concerened with applied

engineering. Doctor of Engineering - 1936 : note of Univ. Michigan.

http://pesn.com/2005/11/16/9600203_New_Nazi_Bell/

Gabriel Kron and the Negative Resistor

Kron was never permitted to release how he made his negative resistor, but did state that, when placed in

the Network Analyzer, the generator could be disconnected because the negative resistor would power the

circuit.

http://www.icehouse.net/john1/

Just a few days before the onset of his fatal illness, He told his wife,

"My work is so beautiful that I wouldn't give it away for a million dollars."

He died, after a short illness, on March 25, 1968

http://en.wikipedia.org/wiki/LGM-25C_Titan_II

· Chapter seven

· Edwin Vincent Gray (1925-1989)

His motor demonstrated an output of 10 HP(7460 watts of mechanical energy) for the extremely low

electrical input of 26.8watts. This is an apparent energy gain of 278 times the input! This left the Cal-Tec

scientists very uncomfortable.

Edwin Gray was born in Washington, DC in 1925. He was one of 14 children.

At age eleven, he became interested in the emerging field of electronics, when he watched some of the first

demonstrations of primitive radar being tested across the Potomac River.

He left home at 15 and joined the Army, but was quickly discharged for being under age.

At 18 he joined the Navy and served three years of combat duty in the Pacific.

He narrowly escaped death when a bomb exploded on his ship's deck during an attack.

He received an honorable medical discharge after spending some time ina navel hospital with head

injuries. After World War 2, he married his first wife, Geraldine, and started a family in Maryland.

He worked as an auto-body and fender repair man. In 1956 he moved his family to Venice, California.

A few months later he moved to Santa Monica where he began his first business named "Broadway

Collision".

33

A couple of years later, he opened a second shop in West Los Angeles.

Both locations failed early in1960 due to an economic downturn.

He relocated to Prescott Arizona, and then to Littleton, Colorado in 1961.

From 1962 until 1964, he worked in Las Vegas, Nevada, always in the auto-body repair business.

By 1965, Gray relocated to southern California again, and established a partnership with George Watson.

Watson was a master car painter with an established clientele of Hollywood celebrities.

A new location was established in Van Nuys, California on Calvert Street called "The Body Shop".

It was a one-stop, high-end custom auto body& painting shop.

This business prospered well for the next three years until a conflict of romantic interests ended his first

marriage (with seven children) in early 1968.

A divorce followed in 1969.(In 1971, Gray married Renate Lenz, the daughter of Fritz Lenz. They had

three children.

This relationship lasted 7 years. Gray married three more times after that.)Towards the end of 1969, Gray

terminated his auto-body business, never to practice it again.

He sold 2/3rds of the Van Nuys building to his nephew and re-outfitted the remaining portion to build and

promote his next business enterprise.

Somehow, Ed Gray had made a sudden and dramatic shift from the auto-body business to an independent

inventor with an extraordinary technology, with hardly any previous background in electronics.

Members of his family are still baffled by the quick transition. Some say their father was occasionally struck with flashes of profound inspiration. Other researchers say that Gray must have been working

secretly on the motors for years, but family members dispute this.

Gray himself told one of his partners that he received this information from a Russian immigrant named Dr.

Popov, who had gotten it from Nikola Tesla.

But again, family members claim no knowledge of these supposed

events.

While there are similarities between Gray's technology from 1970 and Tesla's "Method of Conversion"

technology from 1893, there is no known lineage to trace the connection between these two processes.

No one ever saw Gray studying the work of Tesla, or running any preliminary experiments.

No one who is still alive, who was associated with these events, knows where the technology came from or

how it developed.

In 1971, Gray formed a limited partnership named EVGRAY Enterprises, LTD. By1972, Gray had

gathered enough investment and development expertise to build a motor.

This unit was submitted to Crosby Research Institute for evaluation at Cal-Tech. Crosby Research Institute

was owned by Bing Crosby and run by his brother, Larry Crosby.

34

This motor demonstrated an output of 10 HP(7460 watts of mechanical energy) for the extremely low

electrical input of 26.8watts.

This is an apparent energy gain of 278 times the input! This left the Cal-Tec scientists very uncomfortable.

The report states the motor operated at "over 99%efficiency", but the rest of the data is a little confusing.

One of the closing remarks says "The EMA system will alter the future concepts of energy use."

On the strength of this report, Bing Crosby came on board as a major investor.

Solid 'Boot' Mallory, of the Mallory Electric Company, who made the high voltage ignition coils used in

Gray's circuits.

By early 1973, EVGRAY Enterprises, Inc. had completed a 100 HP prototype motor called the EMA4-E2.

15 private investors were now involved.

Ed Gray also received a "Certificate of Merit" from Ronald Reagan, then Governor of California, during

this period.

By the summer of 1973, Gray was doing demonstrations of his technology and receiving some very

positive press. Later that year, Gray teamed up with automobile designer Paul M. Lewis, to build the first

fuel-less, electric car in America.

But trouble was brewing when a disgruntled ex-employee made a series of unfounded complaints to the

local authorities.

On July 22, 1974, the Los Angeles District Attorney's Office raided the office and shop of EVGRAY

Enterprises, and confiscated all of their business records and working prototypes.

For 8 months, the DA tried to get Gray's stockholders to file charges against him, but none would.

Since he only had 15 investors, many of the SEC regulations did not apply.

By March 1976, Gray pleaded guilty to two minor SEC violations, was fined, and the case closed. After

this investigation ended, the DA's office never returned any of his working prototypes.

In spite of these troubles, a number of god things were happening. His first U.S.Patent, on the motor

design, issued in June of 1975, and by February 1976, Gray was nominated for "Inventor of the Year" by

the Los Angeles Patent Attorney's Association, for "discovering and proving a new form of electric

power".

Despite this support, Gray kept a much lower profile after this time.

But there were also other set-backs. Paul Lewis pulled out of his deal with Gray in1975 when Gray

couldn't deliver a production motor for Lewis's Fascination car.

Gray made a last ditch effort to secure the needed capital to get his motor into production by calling a press

conference in 1976 and demonstrating his nearly complete, second generation 100 HP motor, the EMA-6.

Unfortunately, this event didn't secure any additional funds for the company.

Shortly thereafter, Bing Crosby died in 1977, followed by 'Boot' Mallory in 1978. This left Gray without

his two strongest supporters.

In 1979 Gray reorganized himself into ZETEX, Inc. and EVGRAY Enterprises, Inc. ceased to exist.

35

In the process of this corporate restructuring, all of his earlier stockholders lost all of their money.

Gray then moved his development operations to Kalona, Iowa where new investors were supporting his research.

This working relationship also failed when these new partners attempted a hostile take over.

In a sudden midnight flight, in the middle of winter, Gray loaded up the technology with all his belongings

and headed to San Diego, CA where stayed for 18 months.

In 1982, he relocated his operations to Canyon Country, California where he hired three assistants to help

build several large demonstration carts.

After a year of work, Gray got suspicious of the loyalty of his employees.

He abruptly fired all of them when they reported for work one morning.

He then moved to a second location in Canyon Country and continued with the construction until early

1984.

Later that year, he moved his operation back to Las Vegas where he stayed till the spring of 1985.

In the summer of that year, he moved to the almost abandoned town of Council, ID (population of 816)

where his oldest son 'Eddie' had settled down.

In Council, Gray finished up the construction of five different motor prototypes and several other kinds of

demonstration equipment.

He then began to produce promotional videos and invited local TV stations to report on his work.

Gray then sought out the services of a Wild Cat oil exploration lawyer and found Mr. Joe Gordon of Texas

doing work in Montana.

The two men formed a partnership under Mr. Gordon's established business Western States Oil.

They also established a branch holding company in the Cayman Islands from which to sell stock in the new

venture.

Gray decided to move again, this time to Grand Prairie, Texas to improve his exposure to international

investors.

On the strength of his videos alone, the Cayman Island operation was selling stock and raising capital

quickly.

An investors from Israel convinced Gray to spend two weeks in the Holy Land were a series of emotional

group negotiations took place.

An agreement was never reached.

They conceded that the technology held a lot of promise, but it was not mature enough to be immediately

employed on the battle field.

In addition Gray insisted on maintaining a controlling interest in what ever deal was cut.

36

For whatever reasons, Gray came back with a much different attitude.

Meanwhile the agents who had been selling his stock in the Cayman Islands decided to give themselves

large omissions, plus whatever other funds they had control of, and quickly move to Israel themselves.

Apparently, they had also oversold the original stock issue by about three times.

Feeling swindled himself, Gray made a final, desperate attempt to get proper recognition for his achievements.

He actually wrote letters to every member of Congress, Senators and Representatives, as well as to the

President, Vice President, and every member of the Cabinet, offering the US Government his technology

for Reagan's "Star Wars" program.

Remarkably, in response to this letter writing campaign, Gray did not receive a single reply or even an

acknowledgment!

In 1987, a person named Reznor Orr presented himself, claiming to be a" Government Contact".

Mr. Orr first made straight forward offers to buy all of Gray's technology outright for a modest price.

These initial proposals did not meet with Gray's approval, and he turned them all down.

At about this same time, Gray's income stream from the Cayman Islands stopped.

Mr. Orr's next offers were much less friendly, and mixed with certain veiled threats.

When Mr. Orr left town, to let Mr. Gray think about it", Gray realized he had a serious problem.

Out of money and under threat, he quickly held a massive liquidation sale, including personal belongings

and family furniture he had had for years.

Only the equipment and materials he could stuff into his Ford F-700 box van were spared. Gray drove to

Portland, Oregon and hid out for six months.

Some time during 1987 - 1988, Gray became ill with a serious case of Pneumonia and was hospitalized. He

had been a heavy smoker all his life.

He never fully recovered from this illness and required Oxygen from this point on.

His reduced lung capacity made it much more difficult to continue his work.

From Portland he moved to Sparks, Nevada. Gray rented a combination living quarters and shop space in a

light industrial area.

He unloaded his truck and began to disassemble all of his demonstration carts.

He was living with Dorothy McKellips at the time who claims that Gray still did experiments during the

day but in the evening all the components were once again taken apart and mixed with other parts.

Early, one morning in April of 1989, about 2:00 AM, somebody suddenly started banging hard on one of

the shop windows.

Gray, in his compromised health condition, got out his gun and went down stairs to frighten off the intruder

with a warning shot.

37

The gun failed to fire. A few minutes later, Dorothy found Ed on the floor.

It is presumed that the resulting stress caused Gray to suffer a fatal heart attack, although the exact cause of

death was never determined.

He was 64. The identity of the late night visitor is not known.

Gray's oldest son "Eddie" flew to Sparks, Nevada to identify his father's body.

Later, he spent several months attempting to help a Kansas group recover the technology.

But, Dorothy would not release any of Gray's equipment until she had received a large payment for herself.

The Kansas group then got a court order to take possession of the technology.

But the document was poorly worded and did not define exactly what "technology" really meant. The order

did state that they had rights to all of the motors.

Dorothy caught this fact and gave them just the bare motors, keeping all the power converters and other

things in her possession.

Dorothy then decided to have the last laugh before this looming legal battle could escalate much further.

She had all the remaining equipment, videos, parts, drawings, and laboratory notes hauled away and

dumped in the local land fill.

Apparently none of the remaining systems that the Kansas group had on hand were complete enough to

reconstruct.

Meanwhile, the remaining millions of dollars of investor capital in the Cayman Islands bank account were

tainted by the fraud of the over-sale of the stock.

Ultimately, these funds were either confiscated by the local government in fines or simply swallowed by

the bank, since no one could withdraw the funds without being arrested.

This account of the life and times of Edwin V. Gray was compiled by Mark McKay, of Spokane,

Washington, after numerous interviews with a number of Ed Gray's surviving children.

This account is an attempt to piece together the most accurate retelling of Ed Gray's story ever made

available to the public.

Many of the details in this account are in direct contradiction of earlier accounts as reported in the newspaper clippings from the 1970's.

These earlier accounts should now be considered to be in error.

NOTE: For builders, Edwin Gray used a TRIAC in one of his designs. These devices can be purchased in

the single dollar range. They also have 2 types of semiconductor material in them. PNP and NPN. These

have variations in switching speed. This may be useful in setting up the conditions that will help to produce

what's needed in a Velocity power source.

TRIAC, from Triode for Alternating Current, is a genericized tradename for an electronic component that

can conduct current in either direction when it is triggered (turned on), and is formally called a bidirectional

triode thyristor or bilateral triode thyristor.(wiki)

38

· Chapter eight

· Dr. Henry Moray (1892-1974)

The genius from Utah

Dr. Henry Moray, an early inventor who was able to get it done with free energy, without quite knowing

how he did it and asking himself the question "where is the free energy was coming from?" A quite noted

situation when we are talking about free energy. A quite spectacular man was Henry Moray.

AS a younger man He studied at Uppsala University in Sweden, about 45 miles north of Stockholm.

He had designed and built electrical generating stations.

Moray developed what he termed the "Moray Valve"—a device for extracting "radiant energy" from the

universe.

Dr Moray had spent years building crystal radios using many different crystals.

He found that some from Sweden had a particularly affective at amplifying his radio speakers.

The sound could be so intensified as to cause a speaker to fail from to high amount of energy level he was

able to induce into it.

So his early work building radios demonstrated to him that there was energy in our surroundings to power

radios, and can we wonder when it occurred to him that if a speaker could be amplified using the

surroundings, how about electricity itself, could that be done also?

In his own words: "I started my experiments with the taking of electricity from the ground, as I termed it,

during the summer of 1909.

By fall of 1910 I had sufficient power to operate a small electrical device, and I made a demonstration of

my idea to two friends... This demonstration in the early stages consisted of operating a miniature arc

light... It soon became evident that the energy was not static and that the static of the universe would be of

no assistance to me in obtaining the power I was seeking...

During the Christmas Holidays of 1911, I began to fully realize that the energy I was working with was not

of a static nature, but of an oscillating nature.

Further I realized that the energy was not coming out of the earth, but instead was coming to the earth from some outside source.

These electrical oscillations in the form of waves were not simple oscillations, but were surgings --- like the

waves of the sea --- coming to the earth continually, more in the daytime than at night, but always coming

in vibrations from the reservoir of colossal energy out there in space.

By this time I was able to obtain enough power to light the old 16-candlepower carbon lamp for about one

half capacity, and I did not seem to make any further improvement until the spring of 1925."

http://en.wikipedia.org/wiki/Thomas_Henry_Moray

~~~~~

Nicola Tesla had shared the same thinking as to the lack of utility of a static field would be to the

production of free energy.

39

Classical engineering previously accepted that only a kinetic field could produce motive force to drive a

machine.

WE under stand today that the static field cannot only produce free energy; it is the strongest and most

power field of them all.

It combines all of the fields `that comprise our gravity and maintains everything in its original position as

we sail through the Cosmos.

From reading of his observations when his machines were running, he identified seven distinct zones of

operation.

While we're still in the early stages of developing Velocity power sources, could Dr Moray have tapped

into 7 of the 12 static fields of our planet?

While it is from reading research notes, 12 static fields are claimed to be in existence, with reports that

some researchers have already found 9 of these planetary inertial fields.

Standing waves; Static fields, planetary inertial fields, zero point fields, "it's the same thing!"

In 1887 the Michelson-Morley experiment was specifically searching to discover the aether, that has been

posited from the time of Aristotle.

They detected no Aether and returned with a null result.

Electrical particles are considered to have no mass. Zero point field.

Do They show no mass at all? How then can they apply force to drive a machine?

That is where the dimensional aspects of over unity free energy make their appearance in measuring mass

at the plane of the dimension..

If we went sky diving and held a heavy weight in our hands, if we tried to measure it while coming to earth,

it should register nothing, as it is falling and in motion at the same speed as we are.

When the parachute is opened, the weight would show a measurement because there would be a difference

in speed between the weight and the parachute that is braking our ascent.

This is a somewhat analogy of zero point fields.

If everything is in the same relative speed as everything else is, then what could it show other than the

same relative mass as is equal to the same amount of energy with out a torque point to measure it from?

How could a differential value be measured if the measurement point is in the same motion?

Returning to the exciting things that Dr. Moray had worked on. He was able to build his machine to

produce over 50,000 watts of electrical energy. That's 50 kilowatts;a 50kw.

At today's prices of about ten cents a kw, that equals about $120 dollars of free energy a day.

Per year: $43,800.00 dollars worth of electricity.

In horsepower terms, 50kw is multiplied by 1.341 to give us a little bit over 67.horsepower.

In an ordinary vehicle running along level highway at 55 miles per hour, the horsepower to maintain a

constant speed should be in the range of 5 to15 horsepower.

40

This is mentioned, as is it likely that Dr. Moray's over unity free energy machine could have powered cars

and trucks as early as 1928?

Nicola Tesla built an electric car in the 1930s. It used a 40 horsepower motor which he drove it for a period

of time, to allow others to see and ride in it.

He used ordinary vacuum tubes acquired from local retail electronic stores. He used 12 vacuum tubes to

drive the 40 horsepower electric motor.

So both Tesla and Dr. Moray were able to build in the early 20th century over unity free energy motor and

heater drivers, and yet is it curious that both of them at that time didn't consider that the energy could

possibly be coming from the static fields of our planet?

While Dr. Moray never revealed one secret component of his machine to anyone, can we wonder if it was

an early version of a transistor?

He had used spodumene crystals that he collected from Sweden when on his research adventures.

He also used germanium diodes in his machine.

Could we consider that he was so far ahead of his time that there was no theory of operation known that

would explain what was going on in the machine?

He had hoped for using his technology to hitch it to large-scale city sized water pumps to irrigate the desert

and feed the people.

Is it curious that a man of humane mind should be one that peers far ahead and sees what others may have

missed?

Is it even odder and stranger that such an advanced technology should be blunted, impaired and blocked

and not brought into commercial operation?

He told a reporter that he suspected that his machine was operating due to the different sizes of particles?

The question of the kinetic field as the only source to extract energy from shows up in our top thinkers in

this field.

Tesla and Dr. Moray both were of the opinion that it was not possible to extract energy from a static field.

At the beginning, this was my thinking also.

Then putting a scientific calculator to work, checked all of the electrical fields in our world; it was apparent that there was not enough energy to power up any machine except a very small power device using known

kinetic fields associated with our planet.

After thinking about it it was obvious that the only place high power energy could exist was in the static

field.

That was were it had to be coming from. That's when the question of how to extract it came about.

Dr. Paul Hill (1909-1990) wrote: Unconventional flying objects; a scientific analysis.

Whilst 'Unconventional Flying Objects' was well received by pro-UFO commentators and is widely

referenced by pro-UFO writers, it must be noted that the extraterrestrial hypothesis is not generally

41

accepted and that the mainstream scientific view remains that UFO reports contain nothing of scientific

value.

http://en.wikipedia.org/wiki/Paul_R._Hill

~~~~~~~~~~

It was from exploring Dr Hill's book that helped to determine that the energy found in over unity machines

such as Dr. Moray's, was coming from the static field based upon a few considerations.

Dr Hill gave numbers of UFO power supplies that apparently he was privy to and had gained knowledge of

as an insider in the space program.

In the early 1990's a program hosted by Robert Stack (1919-2003) Unsolved Mysteries, did a report on the

UFOs that had visited Belgium.

The Belgian UFO wave peaked with the events of the night of 30/31 March 1990.

On that night unknown objects were tracked on radar, chased by two Belgian Air Force F-16's,

photographed, and were sighted by an estimated 13,500 people on the ground – 2,600 of whom filed

written statements describing in detail what they had seen.

~~~~~~~~~~~~~~~

One of the Belgian F-16' went in pursuit of a UFO.

Using videotape, for about the next 6 months while playing the videotape back did the numbers on it.

The conclusion was that the UFO was accelerating away from the pursuing F-16 as if it were operating

with a power source of about 3.5 megawatts of electrical energy.

How much is 3.5 megawatts?

Lets turn it into dollars and cents, so that it is easier to understand.

3.5 million watts, is 3,500,000 watts.

We pay for our electricity based on kilowatts used, and a kilowatt is 1,000 watts.

So divide 3.5 million watts by 1,000 and that equals: 3,500 kw and at .10 cents per kw, does that come out

to $350.00 dollars per hour of operation?

Now for horsepower, the UFO demonstrated; taking our 1.341 times our kw of 3,500 and it shows: 4693.5

horsepower developed by the engines that powered the UFO.

Dr. Hill presented optical spectrum analysis of the power source driving UFO's and from what this writer

could determine from them is that the extraterrestrial craft he used to gauge his numbers, produced about

80 megawatts of power.

In ratings, that's over 100,000 horsepower. About 107,280 hp.

Another report using optical spectrum analysis indicating that a craft sitting and hovering showed 2

megawatts of power.

So if we are close, when the UFO moved out of range of the Belgian air force F-16, rather than putting the

pedal to the metal, was it just a light tapping of the gas?

3.5 megawatts and the craft was able to deliver 80 megawatts.

42

This was about a 30-foot diameter craft.

So think for a moment: How much would it produce in retail electrical energy if it was sitting in the back

yard hooked up to our power lines?

Using our .10 cents per kw, 80,000 kw equals: $8,000.00 dollars per hour of operation?

If we have an extra 30 feet in our back yard, how much a year is that in free energy?

Lets see: $8,000 dollars per hour times 24 hours equals: $192,000.00 dollars a day.

How much is that per year then: 365 x $192,000.00 equals: Seventy million eighty thousand dollars.

That's: $70,080,000.00 a year!

You can check the math, to insure it's correct.

The craft that the kind ones dropped at Roswell in 1947 was a sailboat of the skies.

That is, it had no motor on board of it. It was flying using the energy in the fields of our planet.

In Dr. Moray's 50,000kw machine he used a silver and copper bar to help operate the machine.

Somehow he knew intuitively that they were needed to buildup to the high level energies that his radiant

free energy machine demonstrated.

Michael Faraday was experimenting attempting to tap into earth motion in the 1830s.

He used copper and steel wire.

Can we only imagine what would have developed if he had used copper and silver wire?

Copper and silver are very similar electrically. Both are very good conductors, silver is the best conductor

of the metals. Copper is the next best metal for electrical conductivity.

Heat follows the same. Silver conducts heat better than copper.

In a free energy machine as Dr. Moray built, Silver and copper both have similar electronic configurations

of their outer electron shells, this feature can be used to increase the potential energy that can be extracted

from the circuitry when in operation.

The outer electron shell configuration for copper is 3d 10 ....4s 1.

Silver is 4d 10....5s 1

As you may notice, the 3d of copper to the 4d of silver, and both have 10 electrons in the shell.

And the same with 4s of copper and 5s of silver both have 1 electron in their shells.

These electrons can exchange in both directions. This can provide a quantum increase or decrease in energy

exchange value in the transition from one level to the other.

A suggestion was given that these metals create in an electronic circuit something that is analogous to a

flywheel in a steam pump. They allow the circuitry to develop a higher amount of energy when the circuit is oscillating.

43

Can we thank Dr Moray for introducing us to such grand and eloquent machines such as he was able to

produce early in the 20th century?

Is it not more startling that his discoveries have never been utilized to advance the state of circumstance

that the human race is living in?

Dr. Moray also demonstrated the ability to decontaminate radioactive particles. Placing a piece of

radioactive material in a controlled zone of his machine would make the sample become neutral after a

given period of time.

Once again, he didn't understand how he did it, yet he did it.

Think of riding on a bus with the window slightly open.

You light a piece of incense and the fumes rise up and vent through the open window. The bus is your forth

dimension, as you already have three dimensional motion, and the outside of the bus is the fifth dimension.

Same thing is happening in a velocity power source. Energy can be vented into the background of space.

The energy flow process works in both directions.

Is it curious that there was very little bourse available to fund the research work of Dr. Moray?

Does it seem that everything returns to the question of bourse and who issues it?

Dr. Moray had it all for us way back in 1928. He lit lights and ran motors he rewired to match the

frequencies of his radiant free energy generator.

Of course Nathan Stubblefield did the same thing in the 1890s. We don't want to forget Mr. Paine in 1871

with his magnetic motor that ran a saw.

While reading the other day there was a story that American navy forces were working to do a joint

operation with the Vietnamese navy.

It seems that there has been gas and oil found at some of the Islands in the south China sea and America

and China need to face each other as to who will be able to access these gas and oil deposits.

Is there any chance that someone somewhere is fooling us about somethings?

If we build some velocity power sources, will we need to be drilling and looking for oil resources to power

our machines?

Dr. Moray's machine weighed in at about 60 lbs. How much money do you expect it will cost to build a

machine such as he built in the early days of the last century?

We figured $43,800.00 dollars per year of electricity produced by Dr. Moray's 1928 machine.

With solid-state components of today available, could we expect to purchase such machines for not more

than a few thousands of dollars?

Are we ready to move forward into a new world of plenty for all with clean skies above?

Shall we look a little further? How about a Visit to the work of the Late Dr. Mallove and take a look at

shearing force heaters and cold fusion.

44

· Chapter Nine

· Dr. Eugene Mallove

Shearing force oil heaters

Eugene Franklin Mallove (June 9, 1947 – May 14, 2004) was a science writer, editor and publisher of the

magazine Infinite Energy, founder of the non-profit New Energy Foundation, a strong proponent of cold

fusion, and a supporter of research into that and related fringe science topics.

Mallove authored Fire from Ice, a book detailing the 1989 report of table-top cold fusion from Stanley Pons

and Martin Fleischmann at the University of Utah.

Among other things, the book claims the team did produce "greater-than-unity" output energy in an

experiment successfully replicated on several occasions, but that the results were suppressed through an

organized campaign of ridicule from mainstream physicists, including those studying controlled

thermonuclear fusion, trying to protect their research and funding.

Vacuum energy, Zero Point Energy or "ZPE" for short, aether energy, or space energy.

These are descriptions of vast energy sources from the vacuum state.

Dr. Mallove was the link that brought this writer to uncover the facts that confirmed that there is such a

reality as free energy.

It was The report of the members of the Russian academy of sciences of what they had found after testing a

machine made in Moldavia that produced more energy output than went in.

Dr. Mallove had decades of research into the field of free energy experience with him. Regrettably, Never

met him personally, so it is only his research papers that are left for us to work with.

In the early 1980s he had looked at devices called shearing force oil heaters.

These are machines that in some respects could be compared with a washing machine. That is, there is a

high-speed basket with small holes in it, which swirls at high rotational speed in side a drum, not unlike a

washing machine.

There was a patent issued on a shearing force oil heater device in 1927. It claimed a 7 times over unity

factor. There have been demonstrations of these sorts of machines for many years. The heat from them was

inconsistent, and as that it was not understood what was going on at the theoretical level, it was difficult to

boost the amount of energy output. They've remained curiosities, Interesting anomalies in the field of

engineering as it were.

When Dr. Mallove tested some of these machines in the 1980s he determined that there was indeed an over

unity aspect to them. More energy was coming out than was being put into them.

Challenged by the establishment on his findings, he changed his instrumentation and tested some more.

Again He found small amounts of excess energy.

The amounts were not economically significant and the output of the machines was erratic, so the shearing

for oil heaters never made it to market. The thing that remained interesting if not puzzling about these

shearing force heaters is that they actually did produce a small bit of excess energy.

45

A few years later another engineer researcher James Griggs of Rome Georgia discovered another similar technology using water instead of oil. He patented a shock wave hydro sonic pump in 1993.

He's something about it I found on the web concerning the shock wave pump. :

Excerpt from:

Shock Waves and Steam Heat

by

Richard Milton

For more than two years debate has raged on the Internet about an ordinary-looking metal drum sitting on

the concrete floor of a factory building in Rome, Georgia, 50 miles from Atlanta.

Its inventor, the man about whom the Internet debate is raging, is James Griggs, an industrial heating

engineer.

The invention that has brought Griggs such notoriety is a device that he began developing in 1987, that he

calls the 'Hydrosonic Pump' and that many of his supporters believe is over-unity, in that it generates

around 30 per cent more energy as heat than is put in as electricity.

To the skeptics, the Griggs Gadget is, at best, a case of self-delusion on a grand scale, and, at worst, a case

of scientific fraud.

To his supporters, the pump is the first unequivocal public demonstration of undoubted over-unity.

Jim Griggs told me, 'the pump is based on a theory of what takes place when a shock wave is created in a

fluid.

We know that when you create a shock wave in a liquid there is a minute amount of energy released into the fluid in the form of heat.'

'Most of the previous studies had been done in how to eliminate that shock wave, instead of putting the heat

to a useful purpose.

We've designed a system to take the shock-wave heat energy, capture it, and produce hot water or steam.'

Griggs believes that his device works on perfectly normal principles and violates no laws of physics. Just

what happens when the Hydrosonic pump is filled up with water and switched on is described by overunity

investigator Jed Rothwell who conducted a detailed engineering investigation of the device in January

1994.

'During one of the demonstrations we watched,' he says, 'over a 20 minute period, 4.80 Kilowatt Hours of

electricity was input, and 19,050 BTUs of heat evolved, which equals 5.58 Kilowatt Hours, or 117 per cent

of input.

The actual input to output ratio was even better than this, when you take into account the inefficiencies of

the electric motor.'

Entire article here:

http://www.rexresearch.com/griggs/griggs.htm

46

What Griggs found with producing a high intensity shock wave in water and Mallove found in applying

shearing forces to oil is that over unity is a byproduct of this approach from high-speed accelerations upon the substance that is in the machine. So the demonstration of an over unity reaction is found in both oil and

water, both can be used to collect over unity energy.

The shock wave pump that Mr Griggs invented and presented to the world in 1993, is of a similar type of

machine that the Russian researchers looked at in Moldavia in 2003.

In the Moldavian cavitations pump test, the researchers found that it produced four times over unity. That

is, for every 1 watt of electrical energy in to the machine, it returned 4 watts of heat energy.

Not a radical thing in terms of economics because of the fact that for ever four units of energy that goes

into an electrical generator at the power station, only a little more than 1 unit of energy appears at your

home outlet. That fact that it was over unity, is the thing of interest.

What could be more interesting and exciting than the fact that any over unity energy makes its appearance

in these machines?

Dr. Mallove had researched the issue of cold fusion from when it was first announced up and until his

passing. The announcement of a cold fusion process inspired hope for a new age of clean free energy that

appeared to on the horizon when in March of 1989 Stanley pons and Martin Fleischmann made their

presentation at the university of Utah.

Cold fusion gained attention after reports in 1989 by Stanley Pons and Martin Fleischmann (then one of the

world's leading electrochemists)[2wiki] that their apparatus had produced anomalous heat ("excess heat"),

of a magnitude they asserted would defy explanation except in terms of nuclear processes.

The cold fusion research that Dr. Mallove had spent significant time investigating began in 1989 with the

this announcement of over unity energy production in a cold fusion process.

Pons and Fleischmann had demonstrated an over unity reaction and made the case that it was a cold fusion

reaction that was producing the excess energy that was observed in their results.

It seemed to contradict current knowledge in the field.

Steven E. Koonin of Caltech called the Utah report a result of "the incompetence and delusion of Pons and

Fleischmann," which was met with a standing ovation.[47 wiki]

On April 30, 1989, cold fusion was declared dead by the New York Times. The Times called it a circus the

same day, and the Boston Herald attacked cold fusion the following day.

Cold fusion supporters continued to argue that the evidence for excess heat was strong, and in September

1990 the National Cold Fusion Institute listed 92 groups of researchers from 10 different countries that had

reported corroborating evidence of excess heat.

However, no further DOE nor NSF funding resulted from the panel's recommendation.[55 wiki]

By this point, however, academic consensus had moved decidedly toward labeling cold fusion as a kind of

"pathological science".[8wiki]

Dr. Mallove studied this curiosity of a thing called Cold fusion and wrote a book on it by that same name.

When the Russian academy of science group looked into the Moldavian cavitation pump that demonstrated

excess energy, they specifically wanted to see if there was a cold fusion process happening. This may have

47

shown up if they had found helium or tritium . They found neither when they examined the water that was

in the machine after running it through its tests cycles.

No helium, no tritium, so no fusion or other atomic level process. No radioactive waste, no harmful

byproducts, yet the machine produced 4 times as much energy as went into it?

They also checked for the presence of ash, and found none. There was no ordinary chemical process at

work in the cavitation pump. So where was the free energy showing up from?

So in 2003 after thoroughly testing the Moldavian cavitation pump the Russians show us proof of over

unity, but where is the energy coming from? NO fusion of any type and no ordinary chemical or

combustion reaction, so what can explain it?

In his report, Dr. Lev Saporski, leader of the team that investigated the Moldavian heater suggested that it

may be a zero point reaction that is providing the energy that is producing the excess heat discovered in the

Moldavian cavitation pump.

The zero point field, is actually fields. 12 of them are known, these are what Tesla discovered in 1910 when

lightening struck and his electric coils responded and stored an electrical charge in them . He found three of

them. From looking into his notebook, it looks as if Michael Faraday may have found 2 of them in the

1830s.

When we speak of the static field, it is actually fields. As conventional engineering theory holds that only a kinetic field can provide a source of power, how has this gone unquestioned for so long now?

Did Dr. Mallove read of Dr. Saporskis suggestion that it may be a zero point field that was providing the

energy found that brought him to puzzle it out and then make his suggestion that we were only a few

months away from unlimited clean heat and electricity?

Here's how it can be looked at. Oil or water, or most anything else is comprised of three basic things.

Protons, Neutrons and electrons. That's what most atoms are made of.

The electron is the measuring stick for the others. 1 electron mass unit. 1emu.

The proton is about 1800 emu.

Neutron about the same 1800 emu

The proton and neutron are tightly bound together. The electron orbits and is free to exchanger energy with

nearby electrons. All matter is electrical in nature. Matter is based on atoms. For our purposes, we consider

them to weigh 3600 units. It is the separation of the fields associated with the atom that is proposed as to

why the static field reaction is brought about. The speed variation between the electron, 1 mass unit, and

the nucleus, which is a proton and a neutron, about 3600 mass units. Nucleus and proton tightly bound

together, while the electron freely travels throughout its orbit. For our purposes, we could consider the

electron to be the half spin side, while the nucleus is our full spin side of the circuit.

Removing just the 1 electron (time compression 5th dimension separation), causes a reaction that's 3600 times its original inertial positional stability point. It is 3600 mass units that will attempt to roll, or torque at

the plane of the dimension, with only 1 unit of mass removed from the atom.

Think of the high-speed accelerations of the shearing force oil heaters, or the high speed impacts involved

in the cavitaiton pump, or the design patented in 1993, the shock wave pump.

Think for a moment that when this high speed motion is produced in a cavitation, shock wave or shearing

force heater, how it will affect particles of different mass.

48

The proton neutron is about 3600 versus the electron that is 1 mass unit.

Not unlike a centrifuge, the accelerations in the machines are separating the sides of the electrical fields of

the atom causing the electro-magnetic field separation that is allowing the 3600 magnetic mass side of the

field to begin interacting with the static fields of our planet.

Thinking of an analogy to describe this; how does this sound: Think of a gate with a latch and a pin holding

the gate closed. There are several cows pressing against the gate and they put 3600 pounds of pressure

against it as they are wishing to go to the meadow and enjoy themselves. To open the gate that has

thousands of pounds of pressure against it, you pull the latch pin up with one pound of pressure, releasing

the 3600 pounds that are pressed against it.

To complete the analogy in static field magnetic electrical terms, it is in removing the latch pin (half spin

electron) from holding the gate that releases the cows (full spin magnetic) that activates the 3600 mass units to enter into rotatory motion, and torque against the standing waves of our planet.

So we can extract energy from earth motion with mechanical methods as demonstrated in these shearing

force and cavitation pump devices. While the over unity aspects maybe small, is it not the fact that this has

been demonstrated, really big?

The zero point field, is the static fields. They're the inertial fields of our planet. Same things described with

different names.

These fields are ordinarily considered as having no mass, so how can they provide energy? We've looked

at that with the four elements of free energy. Mass that shows zero in one frame of reference, may show

mass when measured from another frame of reference.

· Chapter ten

· HHO electrochemical gas

HHO gas

HHO is an electrochemical gas that was earlier known as Browns gas.

Many demonstrations have shown the gas to be able to be used in cutting torches that can cut materials that

only very high temperatures should ordinarily be able to accomplish, higher levels than oxygen and

hydrogen are able to do

. The theory presented in this book, Free energy here and now and then, if accepted, may supply the

understanding that the excess energy in these processes, is coming from the motion of our planet and is transmitted by static fields of earth. The energy provided is delivered right at the point of usage; where the

gas flame touches the metal. This is where the static field will be reacting; right at the point of use. This is

proposed as to why such high temperatures can be created in gases that normally do not reach into the very

high temperature zones.

Scientific paradigms and conspiracies

49

Institutions receive funding and have established entire departments dedicated to long established theories,

and so it is argued that these same institutions are ill equipped to challenge their own scientific paradigms

with new theories.

HHO is used to power vehicles and is an over unity electro chemical gas system. The oxygen and hydrogen

value of HHO when its burned in an engine, is found to be higher than can be accounted for by the values

of hydrogen and oxygen? So while the hydrogen is separated from oxygen, the total amount of energy

showing up in the combustion processes is higher than the actual values of hydrogen and oxygen?

How can this be explained? Cold electricity has been demonstrated in hho operation and when its found it

proves that the system is clearly an over unity device and is extracting energy from the static fields of our

planet. An hho demonstration showed 12 amps in a very thin wire connected to an hho bottle. The

experimenter put his meter on it and found 12 amps of direct current flowing. The wire was cool to the touch. The reason this is significant is because with 12 amps in a thin wire, it would likely melt in seconds.

That the wire remained cool is proof of cold electricity, which is associated with the energy being drawn

into the machine from our surroundings rather than being pumped in as is ordinarily the circumstance with

ordinary electricity.

A molecule has a ground state configuration. A charged field can add energy to a molecule and can change

the ground state configuration of the molecule.

This over unity reaction may be what we are seeing demonstrated in hho from these processes. The ground

state configuration has been changed. The charges on the molecule have been rearranged.

As this charged particle is then called upon for a reaction, would we consider that it must first return to its

original ground state configuration to complete a chemical reaction?

If so, then Can we theorize that what is happening is that when the charged molecule, hho for instance, is

called upon for a chemical reaction, in reconfiguring itself, it takes an extension of time to reestablish its

original ground state configuration before it can continue on with the reaction, and in doing so it adds a

time dimension into the reaction.

The time to the completion of the reaction is extended. The charges have to unwind themselves from the

charged position and return to their original ground state configuration to complete the change they were

called upon to do. Planet earth is in motion.

This time delay causes a reaction into the static fields of our planet to maintain the position of the atoms in

the reaction field due to earth motion. What is hypothesized here is that the time lag of the charged particle,

causes a dimensional separation between, the materials undergoing this process, and becomes separated in

time, pouring energy in from earth motion to maintain the original stable position of the material.

So at the base level, whether it be a shock wave cavitation pump, shearing force oil heater or electrical or

catalytic reconfiguration of the ground state of the molecule, as in an hho bottle, all of these process's are

producing time variations between the two sides of the electro-magnetic properties of the material or

devices they're operating in.

Could we say that these are all operating in the range of consideration as velocity power sources?

Can we return to the question as to why more energy can be returned from a process than is put into it?

Think again of the size of our particles. 1 emu for our electron and 3600 emu units for our nucleus of the

atom.

The charged particles would mostly reside on our half spin side of the combined electromagnetic spectrum.

50

The voltage side is our half spin side of the combined fields. When they dislocate from their original

position and accelerate into speed of light ranges, their rate of time flow slows. This opens up the other side of the field (magnetic) to extract energy from the static fields that derive their source from planetary

,motion. The energy shows up in the full spin side that remains in the 4th dimension.

As a thumbnail, a 50% return on energy into and out of a system at its base level could be useful to make

an educated guess at to what the dollars and cents of it look like.

That is, 3600 mass units of the atom are put into action when 1 electron mass unit is displaced into the time

space continuum.

So at a 50% return, a device operating extracting energy from planetary motion through the static fields

could be expected to be about 1800 times as much energy output as energy into the system.

We have the cavitation shock wave pumps that have proven 4 times over the input energy. Edwin Gray's

motor operated at 287 times over input energy.

The various different electrical gases that are used now for welding cutting and brazing show exciting

properties. High temperature cutting, when oxygen and hydrogen going beyond their high temperature

range. Why do they show these properties that are outside of the ordinary parameters of the gases involved?

Does it look as if our planet is delivering energy to them through the static fields? Could it be anything

other than that?

From reading of charged particles as a way to extract energy, it looks as if there are a large number of areas

that it could be effectively employed in.

For instance, consider food processing. One ingredient of a recipe could be energized and then using that charged electrical state combined with another ingredient at a specific point in the process that could allow

it to produce an over unity static field reaction to cook the product.

Think of the cost reductions possible. For every $1800 dollars of energy costs, using the static fields could

it turn it into about $1 energy costs then? When will we be using this type of wondrous technology?

Convinced yet that it's a Here and now technology?

The first to document the use of free energy was Alexander Graham Bell.

We have absolute irrefutable proof of over unity in the 2003 Russian academy of sciences group and the

Moldavian cavitation pump. Mr. Bell was the first in history to give us proof of free energy machine

operation. Does it seem possible that it was in 1855?

He was working as a telegraph operator and did some experiments. He disconnected the battery during a

thunderstorm that was running the telegraph and discovered that he could transmit for a brief time. Then it

would stop. He first guessed that the thunderstorm was putting energy into the air for use.

After a while a charge would build up, and again he could transmit a message using free energy until he

used the charge up. He could repeat this process again and again. Rather than the thunderstorm producing

the free energy, he saw that it was available all of the time.

So while it was a small amount he was extracting from where he did not know, it was there for sure and he

proved it way back in 1855.He tried it again after the thunderstorm had passed and found that it would still do it. So he understood then that it was not energy from the thunderstorm that he was tapping into.

So we have proof positive of free energy now going on about 157 years.

51

In the next and last chapter we'll learn of Bob Boyce and his creation of the hex controller.

· Chapter eleven

· The hex controller

While searching around out here on the web have been looking, reading, listening and viewing all sorts of

different people with a variety of thoughts upon their minds concerning free energy.

There are numerous web sites dedicated to the free energy field.

A researcher in California has created an electronic circuit he named the 'hex controller.' It has shown

excellent ability to charge an ordinary 12-volt battery.

It's been in service doing just that for 3 years now. Here's the promised blurb on it:

Child Rides EV Toy on Boyce Free Energy! - A South African experimenter has modified an electrolysis

circuit developed by Bob Boyce so that now it recharges his daughter's electric vehicle riding toy.

What makes this remarkable is that the energy is not drawn from the wall but from the environment

somehow.

He's done this around 35 times now and knows of three replications by others. (PESN; Nov. 12, 2009

Here's a link to see it in operation:

http://peswiki.com/index.php/Directory:Bob_Boyce_Hex_Controller

What can it be that these researchers are on to that has the scientific world scratching its head and

wondering where this energy is coming from?

From seeing and riding in a vehicle equipped with an hho system, it is real convincing when the accelerator

is pressed and it feels as if the engine has a turbo charger attached to it.. Many people have been building

them and demonstrating them for more than a half a century and more.

They are absolutely the real things. This is what Stanley Meyer was building in his workshop. He was

running an engine totally on water.

It seems that Stanley went to the head of the class in the field. He had a car running on 100% hho.

Japanese engineers have an all water powered car operating. Here is a video at You tube that you can see

this amazing development for yourself:

http://www.youtube.com/watch?v=CrxfMz2eDME

On the inside powering the car is an hho system that operates it. It is absolutely positively an over unity

free energy machine. Is this cheerful? If you have the opportunity, watch this video and see how cheered up

it makes you. It is an hho system that is powering this Japanese water car.

The area of most interest at the personal level for me is motor driver systems.

52

The world suffers for too long now. Water water everywhere, but not a drop to drink. Plenty of salt water

available, yet it needs large amounts of energy to convert it for drinking. These velocity power source machines will give us this ability.

To drive large city sized water pumps takes multi thousand horsepower motors. Consider what will be

needed to drive long distance water pipelines throughout the world. The costs in electrical energy using

conventional fuel sources makes them available only to the richest of societies.

With the use of Velocity power sources, once the system is up and running, the energy cost is not an issue.

Labor, management and maintenance are the things to budget for.

With this thinking about large motors, would like to talk a little about the circuit called the hex controller.

This is the invention of a fellow named Bob Boyce, who lives in California.

Here's the story, as I understand it from listening to his interview.

He was operating an electronics business in Florida and building hho units for a boat racing team.

HHO units add horsepower. There is a circuit that is used on the systems that pulses direct current into the

bottle. While working on a hho system with a car on jack stands running the engine. When a certain throttle

position was reached, the engine jumped alive as if a turbo charger had jumped in.

It was puzzling what could have been going on. Using the oscilloscope and examining the signal into the

hho bottle, discovered an alternating current signal mixed in with the direct current signal that was in the

bottle.

Not knowing where the extra energy was coming from, he called it the "anomaly." He spent the time and

effort to create a circuit board that combines both a pulsed DC signal, and a signal like the AC that he accidentally discovered.

Now I've never seen a schematic of this hex controller, so I'm not trying to put words into anybody's

mouth. This is my interpretation of what it is all about.

Checked the parts list and took a general picture of what it is that may be going on here.

What may have been discovered, was that the alternator on the car that was powering the hho system, had a

intermittent or leaky diode in one of the field windings.

An ordinary 12-volt alternator on a car has 3 windings with 2 diodes on each. These convert alternating

current into direct current for use by the systems of the vehicle

How lucky are we that there happened to have been a circuit in the alternator that had one diode that was

accidentally sending AC mixed into the DC and sending them both into the hho system?

Again, not trying to describe this circuit having never worked on one personally, when we have a schematic

we can explain it more thoroughly.

Yet a DC pulsed system with an AC pulse riding with it is the basic needed type of signaling to power up

the circuitry for a motor driver, or most any other type of coil operation that will extract energy from the

static fields of our planet..

In this type of split circuit the DC is the magnetic full spin side. The AC is the half spin, volts side of the

circuit.

53

Here is a link to an interview with the inventor of the hex controller:

http://video.google.com/videoplay?docid=1779100537035350538#

The first solid-state patent issued was in 1915. That's less than a century ago. For all history before that

time, there never were such things as we have seen in our lives.

My grandfather had horses to deliver his barrels in Chicago years ago. Reports I've heard are that he was

also the first guy in the barrel business to have a truck to deliver them.

So the world needs the Velocity power sources. It needed them long ago.

Why have we been halted in our path of progress? What will it take to move forward now and feed the

people; House the people.

Its of course more than a mere technical thing, where is the leadership to guide us into a new tomorrow?

Velocity power sources is a term that was told to Professor Hernandez when he journeyed into space with a

lovely extraterrestrial lady named Lya. Here's a link:

http://rune.galactic.to/lya1.html

Any doubt that they're real? Which one you ask? Both the extraterrestrials and the velocity power sources?

Once we start operating velocity power sources the land will be spirited up. Food and fiber will be made

more available everywhere these machines are sure to go. Environmental destruction as the current way of

life should end.

For those who have the technical interests lets work together and build something that will prove the worth

of these machines. Something that will fund us to move forward and build the systems that will bring abundance to all of Gods kids on this world traveling through the Cosmos; bring us to a new day of

abundance, peace and harmony.

Hope this write is thoughtful enough to share the basics of how these Velocity power sources operate.

Though we cannot normally sense it, and it took about 2500 years to prove it, be assured planet earth and

we are in high speed motion 24 hours a day, seven days a week, 365 days a year, with one added every

four. So is it time to leap to it?

Velocity power sources; that's what they are. Who wants to build them? Hope you're inspired to act up and

help bring velocity power sources online and into our world.

It's a tall order, shut the prisons, stop the war; feed the people. Make time, to have some fun.

Life's short, have fun; be nice.

And now a word from our sponsor:

"Love one another."

"That's it!"

"You're smooching it."

"Have mercy."

54

The End

~Afterwords

Friday, May 25, 2012

Where's the free energy coming from?

Ancient Egyptian bakers used it; modern brewers and bakers still employ its value to add to it. The simple

life of the fungus. So what's it got to do with free energy here and now and then?

Who was it that first noticed that when yeast acts upon wheat water and some salt, it turns into a soon to be

delicious thing we call bread? Bread, the staff of life, what's it have to do with free energy here and now

and then?

Since before the time of the pharaohs our ancestors have been making and baking varieties of bread. Added

flavor, better taste, more energy for life. How do the yeasts add energy to the process? Why has this

question remained unanswered for such a long time?

The question with making bread relates to where is the extra energy coming from that appears when the

yeast takes effect and rises the dough.

The product after the yeast works its magic, is more energy in the dough than before the addition of the

yeast.

So where is the free energy coming from?

Some reports indicate that bread has been made using yeasts possibly for 15,0000 years or more.

IN a quick explanation, the extra energy that we find appearing in the dough after its put to work with the

yeast action; it's coming from the motion of our planet.

How is this happening?

The energy is being ported into the dough by way of our planet's static fields. These are twelve that combine all of our field strengths in together. These fields keep everything in our world in their original

positions.

These are the static, inertial, zero point fields. Before Tesla, and the others who have built over unity free

energy machines, the fungus had already constructed a biological over unity machine and has been tapping

into the free energy from even before the pharaohs and queens.

Mushrooms are funguses, and as with this type of life form, the question has remained as to:" Where is the

energy coming from, the food source for the funguses?" There is no photosynthesis with yeasts.

It is only in the recent era that researchers have uncovered the fact that some enzymes undergo quantum

tunneling.

So, what is quantum tunneling and how does that make a bit of bread grow? How does quantum tunneling

produce free energy?

AS a better question, can we wonder how it happened that the simple fungus was able to create

mechanisms that allowed it to tap into the static fields of our planet? How about we give some thought to

55

how basic life forms organize themselves in such ways as to be able to tap into the free energy zone of our

world?

We'll do this and more as we explore the past and present of free energy machines. Free energy; here and now and then. Velocity power sources. Are you ready to visit the world of the quantum tunneling zone?

The secret of funguses and how they power themselves is not understood in terms of biophysical or

biochemical processes. Well take a look at them in the electrochemical arena and how they achieve their

desire to interact with food sources to produce their motive forces. The driver of the electrical forces that

power funguses and free energy machines we'll need to look into our high-speed planetary motion as we

sail through the Cosmos.

Have you given much thought to how fast it is we are going now?

If the astronomers have it right, 559.4 miles per second combined planet, solar and galactic motion.

To begin our journey, how about we first look at our place in the world that surrounds us, our location

traveling through the universe?

If we are going so fast, where is it that our journey is taking us too? Can we hook some sails up to our high

speed motion so as to power our settlements and have plenty to eat drink and be merry for all of our other

brothers and sisters that are traveling right along with us?

How about a challenge; see if we can discover the route the fungus uses and do the same to extract our

dinner from the high speed motion of our planet. Does that sound like fun?

"Life's short; have fun; be nice." Is there any good reason to be a stranger in paradise?

In real ancient times the Egyptians considered the earth to be flat and standing absolutely still. It was a

settled issue for many thousands of years.

`~~~~

In addition to sugar, yeast also require certain minerals, vitamins and salts for growth.

The mechanisms linking the nutrient environment to transcription and growth rate, however, have remained

unclear. http://www.molbiolcell.org/content/21/1/198.full

Effect Of Yeast (Saccharomyces Cerevisiae1026) Ration Supplementation On Milk Production And Blood

Parameters Of Lactating Baladi Cows

ASHOUR, G.1; HABEEB, A. A.2; http://www.spsa-egy.org/?p=925

Milk energy output was increased ($P<0.05$), while the dry matter intake and gross efficiency of milk energy

production were insignificantly increased due to this supplementation.

~~~

While spending a few days searching for specific numbers, about the best so far as to what will develop

from a yeast looks to be somewhere in the neighborhood of 4-25% increase in output production.

Curiously, in a college biology class in the early 1970s the professor mentioned that it was known that

more energy was found in bread after yeast did their work than where there when they began. It remained a

mystery as to where the excess energy came from. From memory, he never gave a number as to how much

excess energy was found.

56

While working at a major industrial bakery on the South side of Chicago in 1979 a bakers magazine repeated the mention of more energy out than in after mixing yeast and giving it time to rise.

While this is being written, searched out this information. The quantum tunneling effect of enzymes was

detected in 1989. The effects itself of increased energy seen with the use of yeasts had been noted many

decades earlier. Here's a blip and a link.

~~

Quantum Biology Scientific American May 1989

Hydrogen tunneling contributes to an enzyme reaction

It has been known for some years that electron tunneling-a quantum-mechanical effect that enables an

electron to circumvent an energy barrier-has a crucial role in many biological reactions, such as

photosynthesis.

Now researchers at the University of California at Berkeley report that tunneling by hydrogen contributes

to an enzyme-reaction mechanism under biologically relevant conditions.

The enzyme, a yeast alcohol dehydrogenase, speeds up the conversion of benzyl alcohol into

benzaldehyde, a transformation that involves cleaving a hydrate (a hydrogen atom with an electron) from

an alcohol molecule.

The enzyme boosts the rate of this transformation enormously by lowering the energy barrier that must be

surmounted in order for the reaction to take place. A semi-classical model compares the binding of the

hydrogen to the alcohol molecule with the behavior of a mass on a spring. The model predicts that an ordinary hydrogen nucleus-a proton-can hurdle the energy barrier more easily than the heavier isotopes

deuterium and tritium.

This means that the reaction rate for ordinary hydrogen should be faster than that for deuterium and tritium.

How much faster can be calculated precisely. If however, quantum tunneling contributes to the reaction,

the reaction, rate for ordinary hydrogen should be greater than the rate predicted by the semiclassical

model. According to quantum mechanics, a particle's position is uncertain; the probability of finding it at a

given point is smeared out in space.

Tunneling can occur if the region of uncertainty extends to the other side of an energy barrier. Because

particles with smaller mass have a greater uncertainty in their position, they have a higher probability of

tunneling. In biological molecules, electrons tunnel readily across distances of tens of angstroms, whereas a

proton should tunnel less than one angstron-

The heavier hydrogen isotopes, deuterium and tritium, are even less likely to tunnel. To see if tunneling is a

factor in the reaction rate.

Yuan Cha, Christopher J. Murray and Judith P. Klinman prepared two versions of the alcohol, with specific

sites on the molecule occupied by an ordinary hydrogen and a tritium in one, and a deuterium and a tritium

in the other.

During the reaction a benzyl alcohol molecule loses a hydrogen atom to a molecule of nicotinamide adenine dinucleotide (NAD). The workers determined how much of each isotope became bound to NAD.

As they report in a recent issue of Science, the rate at which ordinary hydrogen was transferred was greater

than the rate predicted semi-classically, which would indicate that the reaction is assisted by tunneling.

What is more, the enzyme may facilitate tunneling not only by lowering the energy barrier but also by

narrowing it.

57

This could occur if the enzyme brings the active sites on the NAD and the alcohol very close together. "the

next step is to see what happens if the molecules are kept farther apart," Cha notes. The observation of

hydrogen tunneling could have wide implications.http://www.dhushara.com/book/quantcos/zenq/zenoq.htm

~~~

So how does this thing called quantum tunneling bring energy into our process?

When speaking of things of this nature, does it seem that maybe we need to translate it so that it becomes

understandable? Why is it that the yeasts and other funguses never graduated college, and yet they are able

to quantum tunnel with ease?

Lets start with the big question that has stood for many decades as to why we cannot have free energy.

That question goes something like this:

The first and second laws of thermodynamics tell us that you cannot have more energy coming out of a

machine than going in.

If that's true, then is it a story of Case closed for free energy?

While we know for certain today that our planet is in motion along with our solar and galactic systems, this

took quite some time to find out. About 2500 years ago it was asked and never solved until Copernicus in

1543 AD. Put mathematics to the question of planetary motion. How about we begin here with motion of

our planet?

We've talked earlier about James Bradley and his aberration of light. So we'll leave it here of now.

Have you read of any of the patent medicines of long ago so often talked about? Most look to be harmless

concoctions of herbs and other things, though some have silicates in them. When you think of silicates will

it bring to mind that these are the basis of semiconductors?

Do we have some knowledge now about how the semiconductor can separate the 2 sides of the thing called

electricity and deliver into a system free energy, the energy developed from the motion of our planet?

Could there have actually been something to the patent medicine of long ago?

Wilhelm Reich had a theory about energy and life forces he described as Orgone.

From wiki: Orgone energy is a discredited, fringe science theory originally proposed in the 1930s by

Wilhelm Reich.

Reich, originally part of Sigmund Freud's Vienna circle, extrapolated the Freudian concept of libido first as

a biophysical and later as a universal life force.

The concept of orgone was the result of this work in the psycho-physiology of libido.

After his migration to the US, Reich began to speculate about biological development and evolution, and

then branched out into much broader speculations about the nature of the universe.

This led him to the conception of "bions": self-luminescent sub-cellular vesicles visible he believed were

observable in decaying materials, and presumably present universally. Initially he thought of bions as

electrodynamic or radioactive entities, as had the Ukrainian biologist Alexander Gurwitsch, but later came

to the conclusion that he had discovered an entirely unknown but measurable force, which he then named

58

"orgone",a pseudo-Greek formation probably from org- "impulse, excitement" as in org-asm, plus -one as

in ozone (the Greek neutral participle, virtually

Reich had demonstrated that using a metal screen and placing mercury inside of it would cause a

temperature increase in the mercury. This was proven to be fact. His explanations for it concerned

"convection from the ceiling" would join "air germs" and "Brownian movement" to explain away new

findings."

Understanding the four elements of free energy and the nature of the split electrical fields can we sense

what was happening in this circumstance?

Reich used fiberglass and zinc in building his Orgone boxes. These materials, conductors and semi

conductors, will produce separation in the two sides of the static field. So is it likely that Wilhelm Reich was tapping into planetary motion, knew he had found something, yet didn't have the theoretical basis to

explain it?

http://en.wikipedia.org/wiki/Orgone

Did high speed motion of planet earth help make our bread and cakes and deliver relief into a sore joint or

make a belly ache better using silicates as semiconductors to separate the electrical fields and pour soothing

relief in from the earth's motion?

Until we're certain, is it a best bet yet to: Keep having fun!

You bet!

Nature and Nature's laws lay hid in night: God said, "Let Tesla be", and all was light. - B.A. Behrend

**Nikola Tesla was born "at the stroke of midnight" with lightning striking during a summer storm.
He was born in Smiljani near Gospić, Lika, (the a Military Frontier of Austro-Hungarian Empire, now in Croatia).
The midwife commented, "He'll be a child of the storm," to which his mother replied, "No, of light."**

What follows
here is a listing
of pages about
Nikola Tesla and
therefore
comprise my
tribute to a man
that was very
gifted and
talented, and
unfortunately,
he is all but
forgotten and
almost
completely
unacknowledged
in the
technological
field.

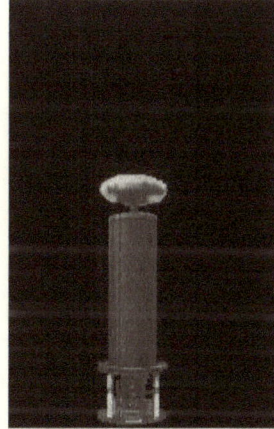

Nikola Tesla was
"a visionary
genius as fertile
as any in the
modern history
of science", and,
had Tesla been
born today, he
would still be
ahead of his
time.

He would be
called a
"genius" by
some, and a
"madman" by
others.

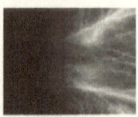

Nikola Tesla was the first to blaze the trail for the creation of incredible, world transforming devices that we, in the world of today, take for granted. Tesla invented such things as: radio, the bladeless turbine, wireless communication, fluorescent lighting, the induction motor, a telephone repeater, the rotating magnetic field principle, the poly-phase alternating current system, alternating current power transmission, Tesla Coil transformer, and more than 700 other patents.

Many of Tesla's patents were in the United States, Britain, and Canada, but many other patents were approved in countries around the globe. Like Thomas Edison, his patents were related to or act as improvements on existing technologies. Unlike Edison, Tesla was more the solitary inventor than the research lab director.

Tesla's inventions and developments include the induction motor, various devices that use rotating magnetic fields, the alternating current poly-phase power distribution system, the fundamental devices of systems of wireless communication (legal priority for the invention of radio), radio frequency oscillators, devices for voltage magnification by standing waves, robotics, logic gates for secure radio frequency communications, devices for x-rays, devices for ionized gases, devices for high field emission, devices for charged particle beams, voltage multiplication circuitry, devices for high voltage discharges, devices for lightning protection, the bladeless turbine, and VTOL (Vertical Takeoff Or Landing) aircraft.

Life® magazine, in a special double-issue, listed NikolaTesla in the "100 Most Important People in the Last 1000 Years". He occupied the 57th position.

Nathan Stubblefield

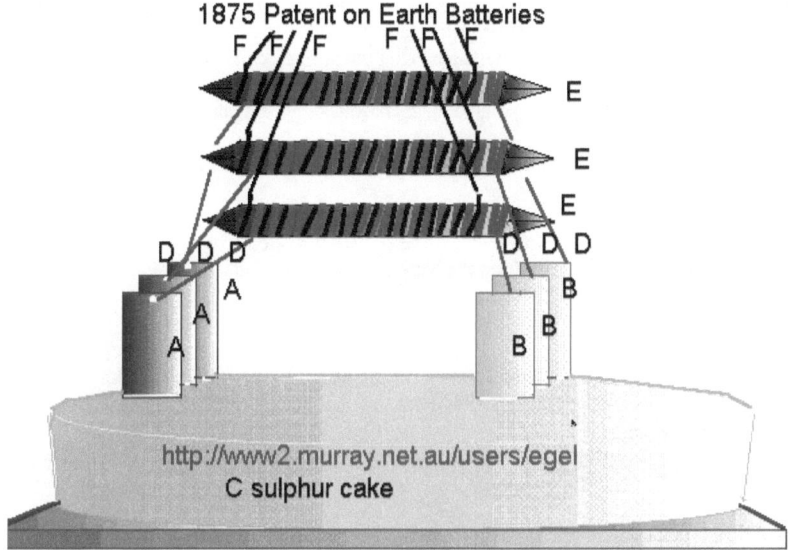

1875 Patent on Earth Batteries

C sulphur cake

http://www2.murray.net.au/users/egel

James C Bryan of Philadephia January 27 1875

1875 Patent on Earth Batteries

There has recently been a question on Earth batteries on the Free energy newsgroup as so many were unaware of the existence of such a device and must admit that I was ignorant of the device myself until I came across this patent and so I reproduce an except from the patent application below.

"The object of my invention is to produce a current of electricity from an earth battery or batteries capable of generating a constant current of considerable intensity to be used for lightning rod and other purposes where voltaic batteries using solutions are now applied.

It is known that if different elements-for instance sheets of zinc and copper-be buried or placed in the earth ,a current of electricity is generated; but I have discovered that if such elements be partly embedded in sulphur so that the dampness of the earth may act in conjunction with the sulphur on the metals, a more intense will be created.

I utilise this in the following way: The current is collected by insulated wires coiled around nickel plated steel magnets ,which are planted north and south in the earth to receive the magnetic current of the earth; a secondary coil or coils of insulated wire surrounds the coil or coils around the magnets and receives by induction, electricity from both the voltaic and magneto-electro batteries.

In the drawing, the voltaic battery is composed of several pieces or plates of chemically pure zinc **B,** and the same number of copper, **A** they are embedded in a cake of sulphur **C** and are connected by a large insulated wire **D**, which being the primary coil between dissimilar elements is extended, without insulation to the base of the sulphur cake **C**, and also in a spiral coil or coils around steel magnets **E**, which are pointed magnetised and nickel plated.

These batteries are planted in the earth north and south, to receive the earths current of electricity according to the magnetic poles.

The primary coil or coils **D** are surrounded by secondary insulated wire F, in a spiral coil or coil to receive by induction ,electricity from the batteries current through **D**

end of patent extract

For those that are interested I recommend you getting a copy of the Borderlands Journal issue volume L111 number one first quarter 1997.

Where there is an interesting article giving more details are other related research into this interesting field..

In the article it is suggested that if you want to try the idea of earth batteries for yourself

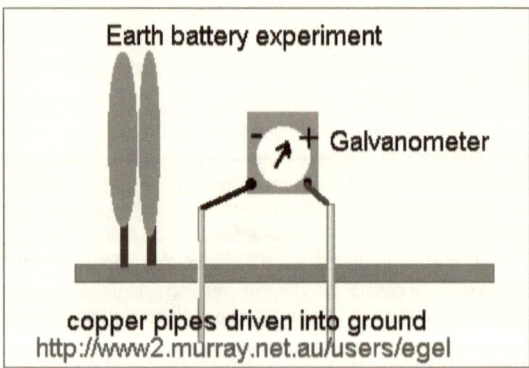

If you want to try it then get two copper rods or pipe and drive them into the ground and then connect a galvanometer(place them near a tree for better results.) apparently the needle always points positive although it may take time for the current and voltage to build up.

It seems that early engineers and telegraph operators knew of the effects of these ground currents when their Edison batteries they had used went flat and long time depleted and their telegraph kept giving out sparks sometimes of greater intensity than when they had been using batteries.

The Stubblefield Earth Battery

Fig one

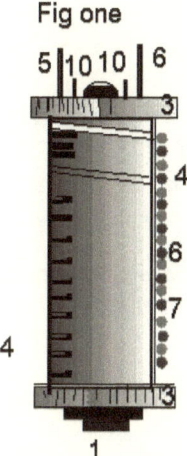

Electrical Battery by

NATHAN B.STUBBLEFIELD OF MURRAY KENTUCKY

The following is a reproduction of Nathan B. Subblefield Patent the drawing may be inaccurate due to some difficulty in reading of the patent drawings Geoff

---

ASSIGNOR OF ONE HALF TO WILLIAM G.LOVE OF SAME PLACE

Specification forming part of letters patent no 600,457 dated march 9 1898

Application filed October 24,1896 serial number 609969 no model

To all whom it may concern.

Be it known that I Nathan B Stubblefield a citizen of the United states of America, residing at Murray in the county of Calloway and

[5]

State of Kentucky have invented a new and useful Electrical Battery of which the following is a specification.

This invention relates to electrical batteries and it has for its objects to provide a

[10]

novel and practical battery for generating electrical currents of sufficient force for practical use, and also providing means for generating not only a constant primary current but also an induced momentary secondary

[15]

current. it is well know that if any voltaic couple be immersed in water placed in moist earth the positive element of the couple will undergo a galvanic action of sufficient intensity

[20]

to produce current when the terminals of the couple are bought in contact , and this form of battery is commonly as the "water Battery ", usually employed for charging electrometers , but are not capable of giving

[25]

any considerable current owing to their great internal resistance . Now the principle involved in this class of batteries is utilised to some extent in carrying out the present invention , but I contemplate, in connection

[30]

with water or moisture as the electrolyte ,t he use of a novel voltaic couple constructed in such a manner as to greatly multiply or increase the electrical output of ordinary voltaic cells , while at the same time producing in

[35]

operation a magnetic field having a sufficiently strong inductive effect to induce a current in a solenoid or secondary coil..To this end the invention contemplates a form of voltaic battery having a magnetic.

[40]

Fig 2

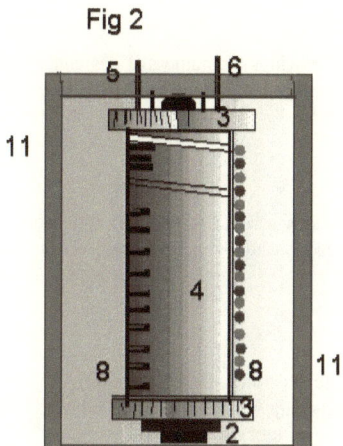

induction properties of sufficient intensity , so as to be capable of utilisation for practical purposes , and in the accomplishment of the results sought for the invention further provides a construction of battery capable of

[45]

producing a current of practically constant electromotive force and being practically free of the rapid polarization common in all galvanic or voltaic batteries . With these and many other objects in view

[50]

the invention ,combination , arrangement of arts herein after more fully described, illustrated and claimed.

In the drawings Figure 1 is a side elevation of an electrical battery constructed in

[55]

accordance with this invention . Fig 2 is a central longitudinal sectional view of the battery , showing the same immersed in water as the electrolyte. Fig 3 is an enlarged sectional view of a portion of the battery ,showing

[60]

more clearly the manner of winding the voltaic couple or in other words , the wires comprising the couple . Fig 4 is a vertical sectional view of the battery , shown modified for use with an induction-coil.

[65]

referring to the accompanying drawings the numeral 1 designates a soft-iron core-piece extending longitudinally of the entire battery and preferably in the form of a bolt having at one end a nut 2 which permits of

[70]

the parts of the battery being readily assembled together and also quite readily taken apart for the purpose of repair , as will be readily understood. The central longitudinally-arranged core-piece 1 of the battery has

[75]

removably fitted on the opposite ends thereof the oppositely - located end heads 3,confining there between the magnetic coil-body 4 of the battery , said heads 3 being of wood or equivalent material. The coil -body 4 of the battery

[80]

is compactly formed by closely-wound coils of a copper and iron wire 5 and 6 ,respectively ,which wires form the electrodes of the voltaic couple , and while necessarily insulated from each other ,so as to have no metallic

[85]

contact ,are preferably wound in the manner clearly illustrated in fig 3 of the drawings.

In the preferred winding of the wires 5 and 6 copper wire 5 is incased in an insulating covering 7,while the iron wire 6

[90]

is a bare or naked wire ,so as to be more exposed to the action of the electrolyte and at the same time to intensify the magnetic field that is created and maintained within and around the coil-body 4 when the battery is in operation

[95]

and producing an electrical current . While the iron wire is preferably bare or naked for the reasons stated , this wire may also be insulated without destroying the operativeness of the battery , and in order to secure

[100]

Fig 3

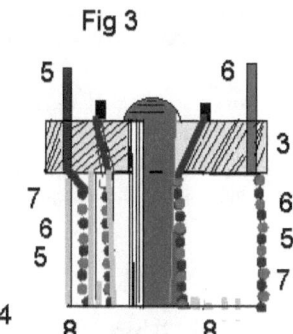

the best results the wires 5 and 6 are wound side by side in each coil or layer of the windings, as clearly shown in FIG 3 of the drawings ,so that in each coil or layer of the windings

[2]

there will be alternate convolutions of the copper and iron wires forming

[5]

the voltaic couple , and it will of course be understood that there may be ant number of separate coils or layers of the wires according to the required size and capacity of the battery. Each coil or layer of the windings

[10]

is separated from the adjacent coils or layers by an interposed layer of cloth or equivalent insulating material 8 and in a similar layer of insulating material 9 also surrounds the longitudinal core-piece 1 to insulate from this core-piece

[15]

the innermost coil or layer of the windings. The terminals 10 of the copper and iron wires 5 and 6 are disconnected so as to preserve the character of the wires as the electrode of the

[20]

voltaic couple; but the other or remaining terminals of the wires are bought into contact through the interposition of any electrical instrument or device with which they may be connected to cause the electric currents

[25]

generated in the coil-body 4 to flow through such instrument or device. In the use of the battery constructed as described the same may be immersed in a cell or jar 11, containing water as the electrolyte

[30]

but it is simply necessary to have the coil-body 4 moist to excite the necessary action for the production of a current in the couple , and it is also the contemplation of the invention to place the battery in moist earth, which

[35]

alone is sufficient to provide the necessary electrolytic influence for producing an electric current. It has been found that by reason of winding the couple of copper and iron wires into a

[40]

coil-body the current traversing the windings of this body will produce a magnetic field within and around the body of sufficiently strong inductive effect for practical utilisation by means of a solenoid or secondary coil 12 as illustrated in Fig 4 of the drawings.The solenoid or

[45]

secondary coil 12 is of an ordinary construction, comprising a wire closely wound into a coil of any desired size on

[50]

Fig 4

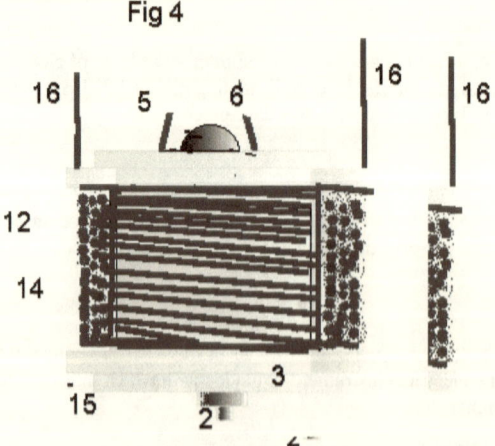

an ordinary spool 13 and increased within a protective covering 14 of mica, celluliod or equivalent material. The spool 13 of the solenoid or secondary coil may be conveniently secured directly on the exterior of the

[55]

coil-body 4 between the heads 3 with a suitable layer or wrapping of insulated material 15, interposed between the spool and the body 4, and the terminals 16 of the solenoid or secondary coil may be connected up with any

[60]

instrument usually operated by a secondary currents- such for instance as a microphone-transmitter or telegraphic relay. The magnetic field produced by the current traversing the coil-body 4 induces a secondary

[65]

current in the solenoid or secondary coil 12 when the ordinary make and break of the primary current produced within the coil 4 is made between the terminals of said coil 4. It will therefore be seen that the construction of the battery illustrated in FIG 4 is practically

[70]

a self generating induction-coil, and it cab be used for every purpose that a coil of this character is used, for as long as the coil-body 4 is wet or damp with moisture electric currents will be produced in the manner

[75]

described. It will also be obvious that by reason of the magnetic inductive properties of the coil-body 4 the core-piece 1 will necessarily be magnetised while a current is going through the body 4 so that the battery

[80]

may be used as a self-generating electromagnet, if so desired, it being observed that to secure this result is simply required connecting the extended terminals of the wires 5 and 6 together after wetting or dampening

[85] the coil-body.

Many other uses of the herein-described battery will suggest themselves to those skilled in the art, and I will have it understood that any changes in the form, proportion and the

[90]

minor details of construction may be resorted to without departing from the principle or sacrificing any of the advantages of this invention.

Having thus described the invention what

[95]

is claimed and desired to be secured by letters Patent is-

1 A combined electrical battery and electomagnet, for use with water as an electrolyte ,comprising a soft -iron core-piece and a

[100]

voltaic couple of copper and iron wires insulated from each other and closely and compactly wound together in separate insulated layers wound together in separate insulated layers to produce a solid coil-body surrounding the soft-iron core-piece, substantially as

[105]

set forth.

2 An electrical battery for use with water as an electrolyte comprising a voltaic couple of insulated copper wire and bare iron wire closely wound into a coil-body substantially

[110]

as described.

3 An electrical battery for use with water as an electrolyte comprising a voltaic couple of insulated copper and bare iron wire wound side by side in separate insulated layers

[115]

to produce a coil-body, substantially as described.

4 An electrical battery ,for use with water as an electrolyte ,comprising a voltaic couple having its separate electrodes insulated from

[120]

each other and closely wound into a compact coil-body forming a self-generating primary coil when moistened and a solenoid or secondary coil fitted on the coil-body of the couple, substantially as set forth.

[125]

In testimony that I claim the foregoing as my own I have hereto affixed my signature in the presence of two witnesses.

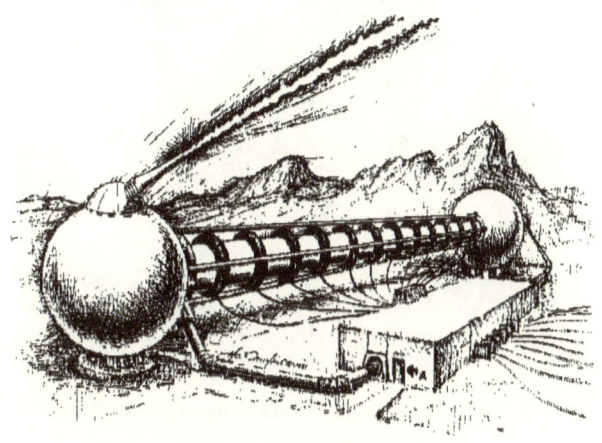

Scalar Potential Interferometer

Figure 6. Multimode Tesla Weapon

In the 1930's Tesla announced other bizarre and terrible weapons: a death ray, a weapon to destroy hundreds or even thousands of aircraft at hundreds of miles range, and his ultimate weapon to end all war -- the Tesla shield, which nothing could penetrate. However, by this time no one any longer paid any real attention to the forgotten great genius. Tesla died in 1943 without ever revealing the secret of these great weapons and inventions.

Unfortunately, today in 1981 the Soviet Union has long since discovered and weaponized the Tesla scalar wave effects. Here we only have time to detail the most powerful of these frightening Tesla weapons -- which Brezhnev undoubtedly was referring to in 1975 when the Soviet side at the SALT talks suddenly suggested limiting the development of new weapons "more frightening than the mind of man had imagined." One of these weapons is the Tesla howitzer recently completed at the Saryshagan missile range and presently considered to be either a high-energy laser or a particle beam weapon, (See Aviation Week & Space Technology, July 28, 1980, p. 48 for an artist's conception.)

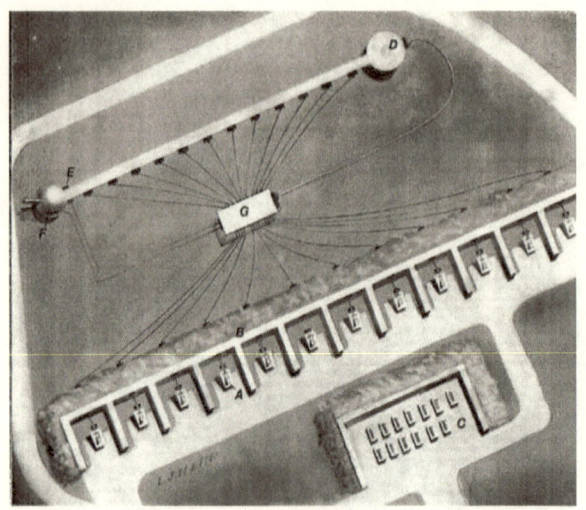

Aviation Week & Space Technology July 28, 1980

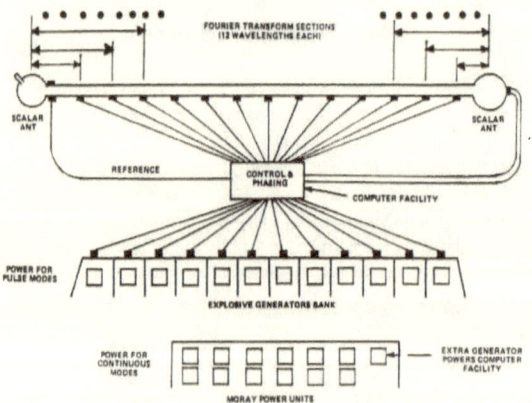

Figure 7. Tesla Weapons at Saryshagan

The Saryshagan howitzer actually is a huge Tesla scalar
interferometer with four modes of operation. One continuous mode
is the Tesla shield, which places a thin, impenetrable hemispherical
shell of energy over a large defended area. The 3-dimensional shell
is created by interfering two Fourier-expansion, 3-dimensional
scalar hemispherical patterns in space so they pair-couple into a

dome-like shell of intense, ordinary electromagnetic energy. The air molecules and atoms in the shell are totally ionized and thus highly excited, giving off intense, glowing light. Anything physical which hits the shell receives an enormous discharge of electrical energy and is instantly vaporized -- it goes pfft! like a bug hitting one of the electrical bug killers now so much in vogue.

If several of these hemispherical shells are concentrically stacked, even the gamma radiation and EMP from a high altitude nuclear explosion above the stack cannot penetrate all the shells due to repetitive absorption and reradiation, and scattering in the layered plasmas.

In the continuous shield mode, the Tesla interferometer is fed by a bank of Moray free energy generators, so that enormous energy is available in the shield. A diagram of the Saryshagan-type Tesla howitzer is shown in figure 7. Hal Crawford's fine drawing of the interferometer end of the Tesla howitzer is shown in figure 6. Hal's exceptional rendition of the Tesla shield produced by the howitzer is shown in figure 8.

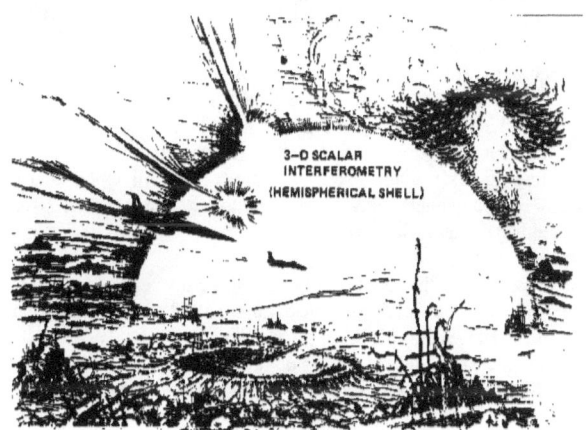

3-D SCALAR
INTERFEROMETRY
(HEMISPHERICAL SHELL)

Figure 8. The Tesla Shield

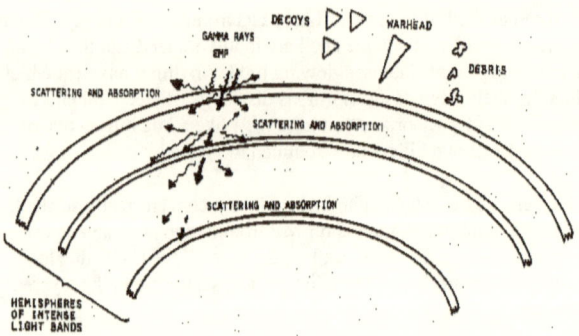

Figure 9. Tesla Terminal Area Defense System

In the pulse mode, a single intense 3-dimensional scalar phi-field pulse form is fired, using two truncated Fourier transforms, each involving several frequencies, to provide the proper 3-dimensional shape (Figure 10). This is why two scalar antennas separated by a baseline are required. After a time delay calculated for the particular target, a second and faster pulse form of the same shape is fired from the interferometer antennas. The second pulse overtakes the first, catching it over the target zone and pair-coupling with it to instantly form a violent EMP of ordinary vector (Hertzian) electromagnetic energy. There is thus no vector transmission loss between the howitzer and the burst. Further, the coupling time is extremely short, and the energy will appear sharply in an "electromagnetic pulse (EMP)" strikingly similar to the 2-pulsed EMP of a nuclear weapon.

This type weapon is what actually caused the mysterious flashes off the southwest coast of Africa, picked up in 1979 and 1980 by Vela satellites. The second flash, e.g., was in the infrared only, with no visible spectrum. Nuclear flashes do not do that, and neither does superlightning, meteorite strikes, meteors, etc. In addition, one of the scientists at the Arecibo Ionospheric Observatory observed a gravitational wave disturbance -- signature of the truncated Fourier pattern and the time-squeezing effect of the Tesla potential wave -- traveling toward the vicinity of the explosion.

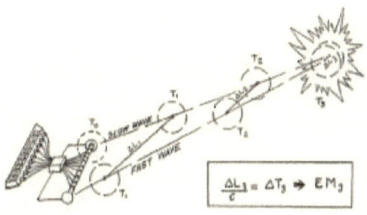

TESLA HOWITZER
(SCALAR INTERFEROMETER)

Figure 10. "Nuclear" Flashes off the Coast of Africa

**Lithuania - 10 Sep 1976 - British European Airways Flight #831
between Moscow and London**

CIA Report Released under FOIA

Figure 11. Continuous Tesla Fireball

The pulse mode may be fed from either Moray generators or -- if
the Moray generators have suffered their anomalous "all fail"
malfunction -- ordinary explosive generators. Thus the Tesla
howitzer can always function in the pulse mode, but it will be

limited in power if the Moray generators fail.

In the continuous mode, two continuous scalar waves are emitted --
one faster than the other -- and they pair-couple into vector energy
at the region where they approach an in-phase condition. In this
mode, the energy in the distant "ball" or geometric region would
appear continuously and be sustained -- and this is Tesla's secret of
wireless transmission of energy at a distance without any losses. It
is also the secret of a "continuous fireball" weapon capable of
destroying hundreds of aircraft or missiles at a distance. An
example of a Soviet test of this mode of operation is shown in
figure 11.

Witness to a super weapon?

**Nick Downie describes the strange lurid glow that flared silently over the
Hindu Kush**

THE SUNDAY TIMES, 17 AUGUST 1980

(Multiple incidents in Sept., 1979)

**TEHERAN, IRAN
17 JUNE 1966**

**SEEN FROM
NEAR
MEHRABAD
AIRPORT**

**OBSERVED 4-5
MINUTES**

**SEEN BY 2
AIRCRAFT**

CIA REPORT RELEASED UNDER FOIA

Figure 12. Tesla EMP Globe

The volume of the Tesla fireball can be vastly expanded to yield a globe which will not vaporize physical vehicles but will deliver an EMP to them to dud their electronics. A test of this mode is shown in figure 12. (See also Gwynne Roberts, "Witness to a Super Weapon?", the London <u>Sunday Times</u>, 17 August 1980 for several other tests of this mode at Saryshagan, seen from Afghanistan by British TV cameraman and former War Correspondent Nick Downie.)

If the Moray generators fail anomalously, then a continuous mode limited in power and range could conceivably be sustained by powering the interferometer from more conventional power-sources such as advanced magnetohydrodynamic generators.

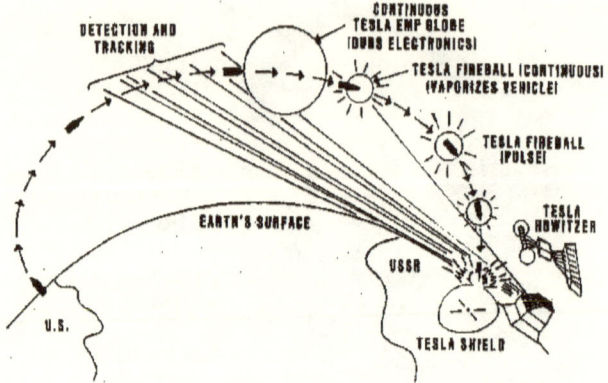

Figure 13. Tesla ABM Defenses

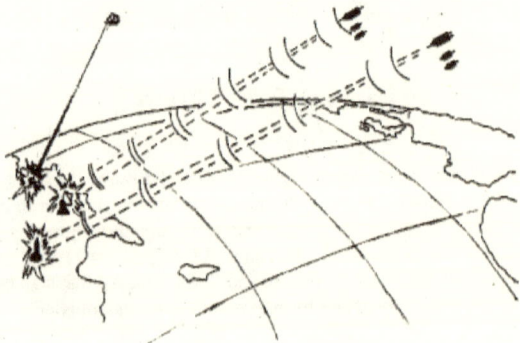

Figure 14. Moray/Tesla Technology: Star Wars Now

Typical strategic ABM uses of Tesla weapons are shown in figure 13. In addition, of course, smaller Tesla howitzer systems for anti-tactical ballistic missile defense of tactical troops and installations could be constituted of more conventional field missile systems using paired or triplet radars, of conventional external appearance, in a scalar interferometer mode.

With Moray generators as power sources and multiply deployed reentry vehicles with scalar antennas and transmitters, ICBM reentry systems now can become long range "blasters" of the target areas, from thousands of kilometers distance (figure 14).

Literally, "Star Wars" is liberated by the Tesla technology. And in air attack, jammers and ECM aircraft now become "Tesla blasters." With the Tesla technology, emitters become primary fighting components of stunning power.

The potential peaceful implications of Tesla waves are also enormous. By utilizing the "time squeeze" effect, one can get antigravity, materialization and dematerialization, transmutation, and mindboggling medical benefits. One can also get subluminal and superluminal communication, see through the earth and through the ocean, etc. The new view of phi-field also provides a unified field theory, higher orders of reality, and a new super-relativity, but detailing these possibilities must wait for another book.

With two cerebral brain halves, the human being also has a Tesla scalar interferometer between his ears. And since the brain and nervous system processes avalanche discharges, it can produce (and detect) scalar Tesla waves to at least a limited degree. Thus a human can sometimes produce anomalous spatiotemporal effects at a distance and through time. This provides an exact mechanism for psychokinesis, levitation, psychic healing, telepathy, precognition, postcognition, remote viewing, etc. It also provides a reason why an individual can detect a "stick" on a radionics or Hieronymus machine (which processes scalar waves), when ordinary detectors detect nothing. Unfortunately there is not room to develop the implications of this human Tesla interferometry in detail, for that must wait for yet another book, presently in its initial stages, that Hal Crawford and I are writing.

Table 5. Orders of Reality

- **PHOTONS ARE**
  - > **PAIR-COUPLED SCALARS**
  - > **VELOCITY-LIMITED TO C**
  - > **CARRIERS OF T**
- **PHOTON INTERACTION**
  - > **IS UBIQUITOUS**
  - > **PRODUCES $\hat{T}$ = C**
  - > **YIELDS 1st ORDER REALITY**
- **SCALAR O WAVES**
  - > **NOT VELOCITY-LIMITED**
  - > **YIELD HIGHER ORDER REALITIES**

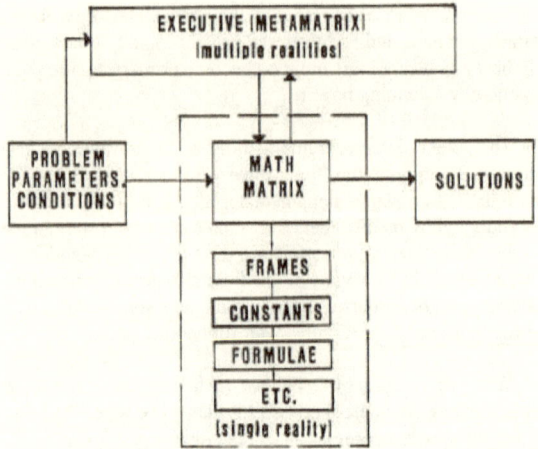

Figure 15. Implications of Tesla Potential

At the July 1981 U.S. Psychotronics Association's Annual Conference in Dayton, Ohio, I presented the first rough paper on the Tesla secret and scalar interferometry. A videotape of the presentation was made and will shortly be available. I am also scheduled to make a special presentation at the Alternate Energy Conference in Toronto, Canada in latter October, 1981. A professional, videotaped two-hour presentation on this subject is also being prepared. Wide distribution of the material through the international underground physics and technology network has already been made. This time, God willing, Tesla's secret will not be suppressed for another 80 years!

And perhaps it is not yet too late. The material has cost me (now) some 16 years of agonizing labor and nearly $100,000 of my own personal funds. No orthodox university, scientific group, foundation, or governmental agency would support such an effort, either financially or otherwise. Indeed, most ordinary journals will not even accept material on such matters. Nonetheless, the area is of overwhelming importance and I truly believe Tesla's lost secret will shortly affect the lives of every human being on earth.

Perhaps with the free and open release of Tesla's secret, the scientific and governmental bureaucracies will be. shocked awake from their slumber, and we can develop defenses before Armageddon occurs. Perhaps there is hope after all -- for even Brezhnev, in his strange July, 1975 proposal to the SALT talks, seemed to reveal a perception that a turning point in war and weaponry may have been reached, and that human imagination is incapable of dealing with the ability to totally engineer reality itself. Having tested the weapons, the Soviets must be aware that the ill-provoked oscillation of timeflow affects the minds and thoughts -- and the very lifestreams and even the collective species unconsciousnesses -- of all lifeforms on earth. They must know that these weapons are two-edged swords, and that the backlash from their use can be far more terrible to the user than was the original effect to his victim.

If we can avoid the Apocalypse, the fantastic secret of Nikola Tesla can be employed to cure and elevate man, not kill him. Tesla's discovery can eventually remove every conceivable external human limitation. If we humans ourselves can elevate our consciousness to properly utilize the Tesla electromagnetics, then Nikola Tesla -- who gave us the electrical twentieth century in the first place -- may yet give us a fantastic new future more shining and glorious than all the great scientists and sages have imagined.

Dr. T. Henry Moray

T. Henry Moray makes some final adjustments in tuning his radiant energy device to tap zero-point vacuum energy

THE MORAY RADIANT ENERGY DEVICE

In the early 1900's, Dr. T. Henry Moray of Salt Lake City produced his first device to tap energy from the metafrequency oscillations of empty space itself. Eventually Moray was able to produce a free energy device weighing sixty pounds and producing 50,000 watts of electricity for several hours. Ironically, although he demonstrated his device repeatedly to scientists and engineers,

Moray was unable to obtain funding to develop the device further into a useable power station that would furnish electrical power on a mass scale.

As a boy, Moray had been deeply inspired by the greatest electrical genius of all time, Nikola Tesla. His imagination was especially fired by Tesla's claims to have knowledge of an energy source greater than ordinary electricity, and by Tesla's emphasis on frequencies as the stuff of the universe. When Moray finished high school in Salt Lake City, he went abroad to study, and took resident examinations for his doctorate in electrical engineering from the University of Uppsala, Sweden, during the period 1912-14. Returning home, his diploma and credentials were interrupted by World War I, and the University mailed him these items in 1918 after the war.

Shortly thereafter, Moray produced his first elementary device that delivered measureable electrical power, and he continued to work diligently on energy devices when he had time. In the 1920's and 1930's he steadily improved his devices, particularly his detector tube, the only real secret of the device according to Moray himself. In his book, *The Sea of Energy in Which the Earth Floats,* Moray presents documented evidence that he invented the first transistor-type valve in 1925, far ahead of the of officially recognized discovery of the transistor. In his free energy detector tube Moray apparently used, inside the tube itself, a variation of this transistor idea—a small rounded pellet of a mixture of triboluminescent zinc, a semiconductor material, and a radioactive or fissile material His patent application (for which a patent has never been granted) was filed on July 13, 1931, long before the advent of the Bell Laboratories' transistor.

Here the Moray radiant energy device is providing free power to 35 100-watt lamps and a 1200-watt iron.

In test after test Moray demonstrated his radiant energy device to electrical engineering professors, congressmen, dignitaries, and a host of other visitors to his laboratory. Once he even took the device several miles out in the country, away from all power lines, to prove that he was not simply tuning in to energy being clandestinely radiated from some other part of his laboratory. Several times he allowed independent investigators to completely disassemble his device and reassemble it, then reactivate it themselves. In all tests, he was successful in demonstrating that the device could produce energy output without any appreciable energy input.

According to exhaustive documentation, no one was ever able to prove that the device was fraudulent or that Moray had not accomplished exactly what he claimed. On the other hand, the records are full of signed statements from physicists, electrical engineers, and scientists who came to the Moray laboratory as doubting Thomases and left with the complete conviction that Moray had indeed succeeded in tapping a universal source of energy that could produce free electrical power.

But in the face of all of this, the U.S. Patent Office refused to grant Moray a patent, first, because his device used a cold cathode in the tubes (the patent examiner asserted it was common knowledge that a heated cathode v as necessary to obtain electrons) and, second, because he failed to identify the source of the energy. All sorts of irrelevant patents and devices were also presented as being infringed upon or duplicated by Moray's work. Each of these objections was patiently answered and nullified by Moray; nonetheless, the patent has still not been issued to this day, although the Morays still keep the patent application current.

One of Moray's efforts to develop the machine involved an association with the Rural Electrification Agency for a short time before World War II. At that time, the R.E.A. was apparently infiltrated by Communist sympathizers and high level officials of a decidedly pinkish tinge. These officials continually urged Moray to turn over all details of his device to the Soviet Union, and even arranged the visit of high-level Soviet scientists to the Moray laboratory to see the device in operation On the initiative of the Communist-infiltrated R.E.A., one person—let us call him "Z"—spent about two months in Moray's laboratory, and succeeded in worming his way into Moray's full confidence. Moray eventually disclosed most of the details of the construction of his special valve to "Z"—the only person in whom Moray ever completely confided.

There is strong reason to believe that "Z" was a Soviet agent, and that this is where the Soviets first obtained the impetus to develop the hyperspace amplifiers later used in their psychotronic weapons.

However, Moray became quite alarmed at the continued attempts of his R.E.A. contacts to get the device into the hands of the Russians. He eventually concluded that he had become involved with a governmental group filled with radicals and reactionaries. Moray became concerned that fifth column activity was actually directed against him in an attempt to steal his device. Quotes from the *Salt Lake Tribune* of December 2, 1941, attributed to Representative Thomas D. Winter, imply that Moray's suspicions of the R.E.A. were well founded, since Representative Winter called for a full-scale investigation of the R.E.A. on the grounds of Communist infiltration. Indeed, Moray was wounded by gunshot in his own laboratory on March 2, 1940, which seems to suggest that his fears and suspicions were based on fact. Moray severed his association with the R.E.A. in February 1941.

However, his basic unit had been destroyed by a hammerslinging witness in 1939; it is not clear whether or not this was the work of "Z" or of someone else. According to his son, John Moray, the man who broke his unit, as well as other interested groups, wanted complete disclosure of the materials and construction — nothing more or less. John Moray, who operates the Research Institute in Salt Lake City, has been trying to continue his father's work since the basic unit was destroyed. Dr. Moray himself died in May 1974.

According to John Moray, highly energetic cosmic rays from space are tapped by the machine, which is in subharmonic resonance with this high-frequency energy level, and it converts this energy level into a usable form of electricity. However, John Moray's use of the term "cosmic ray" is not necessarily the same as that of modern physics, but in fact is the same as the present concept of "zero-point" energy of vacuum. T. Henry Moray envisioned all space filled with tremendously high frequency vibrations carrying vast and incalculable amounts of pure raw energy. He envisioned the dynamic Universe as a turbulent source of great energy, just as the ocean waves carry energy throughout the earth. This was also the vision of Tesla, and after him of Clifford and Einstein, who turned the attention of general relativity to the problem of the nature of the vacuum itself. Clifford and Einstein founded a branch of physics that has come to be known as quantum geometrodynamics, the study of energetic change that occurs in little pieces, including the concept that very small lengths of space, or pure vacuum, themselves oscillate at great frequency and with great energy. In Moray's day relativity was still a strange and unproven branch of physics, suspected and rejected by most of the physicists of the day, and quantum physics was still in the process of being worked out. There was no theory whatsoever predicting that empty space itself not only contained prodigious quantities of energy, but in fact *was* prodigious quantities of energy. But slowly, over the decades, the picture has changed, and the modern followers of quantum geometrodynamics assert the truth of Tesla's original vision. Today we know that one cubic centimeter of pure vacuum contains enough energy to condense into $10^{80} - 10^{120}$ grams of matter! Thus the major part of Moray's thesis—that vacuum itself contains unlimited energy—is vindicated today. In this sense empty space is like a gigantic, restless ocean, and Moray's free energy "tapping" device is no more mysterious than the water wheel. In other words, his thesis that the energy is there to be tapped is correct; it only awaits a practical method to tap it in order to solve the energy problems of mankind forever.

Once tuned in, the power continues and is free for the taking.

Moray thus is vindicated as a man ahead of his time, who simply built a device before any theory existed to explain its operation. Moray met with constant opposition, and his life and that of his family was anything but pleasant. Certain pressure groups constantly tried to force him into selling for almost nothing, or to disclose the secrets of his valve's construction. He was often attacked and he was sometimes shot at on the city streets. In fact, his life was threatened so often that he was forced to install bulletproof glass in his automobile. His laboratory was broken into, some of his components and papers were stolen, and his dogs were continually being killed. Moray was shot in his own laboratory, and he himself always carried a gun. Because he was harassed ceaselessly, over the years he became understandably suspicious and close-mouthed about his work. He sometimes greeted visitors at his desk with a loaded pistol lying on the desk within easy reach, and occasionally confirmed that he would not hesitate to shoot if he were attacked or threatened. One of his: greatest fears was that big interests would take up his invention and simply shelve it to keep it from benefiting the public. When large companies made him offers, he always demanded written guarantees that the device would be put into production and sold to everyone, once it was developed.

Moray also realized the potential of his devise as a weapon. He was fully aware that the only difference between a controlled energy device and an explosion is the rate of release of the energy. According to John Moray, his father received an offer to go to Japan in 1938 because he had built a deathray which operated off the device. According to John, a representative of the Japanese government came to the laboratory and offered a position to Moray, but he refused. Allegedly Moray had managed to rig the device so that it projected a ray of energy in a beam, and could kill mice instantly at over fifty feet. When radiated by the beam, the mice were carbonized immediately. They appeared frozen, but nothing was left except their shells, and they simply fell apart when touched.

In one experiment Moray ran his device for 157 hours without any connection to external power sources, and produced over fifty kilowatts of power during the test. He also found that an additional fifty kilowatts could be added by simply providing another tap further back in the circuit. When he shut the device off from this test, he had proven once and for all that the device was generating electrical energy from free and natural sources without batteries or external power. During this test nothing in the machine heated up; instead, all parts of the circuit ran absolutely sold. This alone is totally unexplainable by ordinary electromagnetic theory, and it strongly implies the truth of Moray's assertion that the device simply collects the energy in each of its stages in a subharmonically resonant manner, in synchronization with the extremely high frequency cosmic oscillations. In other words, since the parts of the machine ran cold, it is clear that the energy was being simultaneously collected at each stage rather than being processed through the individual stages in serial order, since serial processing in the conventional electrical sense would lead to resistance heating of the circuit elements.

Moray's device used twenty-nine stages of his special detector valves, which were difficult to produce, costing about five hundred dollars each. Only about one in four proved suitable for operation. He also had difficulty in obtaining sufficiently pure materials to make his special mixture for the pellet that enabled tube operation in a one-way gating fashion. Moray explained that his device was based on the discovery of a mixture that would act as a one-way gate for the high frequency oscillations of space, so that the energy could go through the material more readily in one direction than another. Thus it was like a one-way gate valve to an ocean wave; the energy "water" could flow in in each valve, but was prevented from flowing back out. The assemblage of multiple stages thus provided a series of collectors which contained enough energy to be useful. Theoretically there was no limit to the number of collectors that could be added, and so there appeared to be no limit to the energy that such a device could produce.